MW00931520

illustrated by @landandworld

BOOK OF INTEGRALS

To the bystander, looking to be entertained by these math scribbles

To the math student, tempted to pick up pens of their own and follow along

To those currently studying the art of integrals and ready to improve their skills

This book is for you

It's happening and I'm so excited. Grab your favorite drink, your favorite snack, put on your favorite music (if that's your thing) and come solve some integrals with me.

Let me ask you something: how do you imagine integrals? Do you just see the weird looking symbol that Leibniz introduced to the world, or through a Riemann sum as the number of rectangles tends to infinity?

$$\int_a^b f(x)\,dx \ \text{ OR } \ \lim_{n\to\infty} \sum_{i=0}^{n} f\left(a + \left(\frac{b-a}{n}\right)i\right)\left(\frac{b-a}{n}\right)$$

i put this on the map!!!

okay?

or maybe your mind immediately goes to the anti derivative of the function *f(x)*

This is all correct, since they obviously are all related. The integral of a function *f(x)* is in fact the anti derivative, and geometrically it represents the area underneath the curve from interval [a,b] where the area is traditionally calculated by finding infinitely many areas of rectangles whose widths are infinitesimally small.

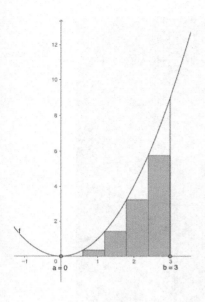
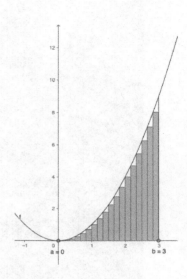

Finding definite integrals (the area underneath the curve) is much easier that finding the general solution to an indefinite integral.

$$\int_a^b f(x)\, dx \quad \text{vs} \quad \int f(x)\, dx$$

We can use various techniques to find the area (definite integral) that lead to our solution, but the indefinite needs much more work, patience, and techniques to uncover the integral's secrets and inner-workings.

Call me crazy, but I relate integrals to the big massive trees I so often see outside my window. Every single one reaches towards the sky, pointing its branches in different directions and has its own story worth telling. These trees require specific environments in order survive. It's our job to go out and discover these requirements.

We're going to do a lot of exploring and uncover the trail of integrals. We're here to examine the beauty of integration, a list of techniques, and understand what's hiding behind these behemoths without making too much of a mess.

We will review common integrals worth memorizing to use in future adventures (which will make our experience more enjoyable). I will show my math scribbles for some of the integrals, but the majority of the time you'll be encouraged to solve them on your own!

Our adventure starts off with the basics and slowly starts increasing in difficulty. I encourage you to take your time and not rush into the book. Alot of the techniques we uncover will be necessary to survive the ones coming after it. At the end of our journey I lay out some species of integrals that show, not only how powerful they are in uncovering our world's secrets, but just how much more there is to study about them.

Living in the Pacific Northwest, surrounded by luscious evergreens, I find myself hiking the trails observing wild and plant life constantly. I'm always in awe at what's out there and what can be discovered. Often times my wife and I will hike very small portions of the Pacific Crest Trail (PCT), a 2650 mile-long trail that traverses through Mexico, California, Oregon, Washington, and finishing at the Canadian border.

It is a treacherous journey when hiking the PCT. Spending months on your own (or with little friends) one can only imagine what hikers encounter and how different of a person they come out from this experience.

This book, my friend, is hiking your very own PCT. Let's see what these integrals have to teach us how we change as a students of the field.

$$\prod_{i} \sum_{n=0}^{\infty} \lim_{b \to \infty} \int_{1}^{b} \cdots dx$$

"lost in the lore of trees my soul sings earthen melodies" - Angie Weiland Croshy

Table of Contents

THE POWER RULE

I really don't think that we need to discuss this very much, but let's see 4 scenarios encountered on trail that requires some basic manipulation, which will lead to the power rule!

$$\int X^n \, dx = \frac{X^{n+1}}{n+1} + C$$

Basic Rule

(A) $\int x^3 \, dx = \boxed{\frac{X^4}{4} + c}$

Radicals — Always Rewrite!

(B) $\int \sqrt[4]{x} \, dx = \int x^{1/4} \, dx$

$\frac{X^{5/4}}{5/4} + C = \boxed{\frac{4}{5} x^{5/4} + c}$

Expanding

(C) $\int (x+1)^2 \, dx$ Expand this!

$= \int x^2 + 2x + 1 \, dx$

$\boxed{= \frac{x^3}{3} + x^2 + X + C}$

One-valve denominator

(D) $\int \frac{x^2 + 3}{\sqrt{x}} \, dx$ → Separate them!

$= \int \frac{x^2}{x^{1/2}} + \frac{3}{x^{1/2}} \, dx$

$= \int x^{3/2} + 3x^{-1/2} \, dx$

$\boxed{= \frac{2X^{5/2}}{5} + 6X^{1/2} + c}$

Notice the progression and how one can change the way the integral looks so that the power rule works!

YOUR TURN!

I used to struggle with these in high school. Mostly because I couldn't add fractions. How silly of me - don't make the same mistake please! I will leave you here to explore and practice these integrals and will catch up with you when you're finished.

Have fun!

1) $\displaystyle\int x^2 + 3x\ dx$

The first integral starts off really easy, just using the power rule. However, the next ones will require you to rewrite radicals into rational exponents, and denominator exponents into negative ones. The real fun has begun, in my opinion!

2) $\displaystyle\int 6\sqrt{x} - \frac{4}{x^3}\ dx$

3) $\displaystyle\int 7\sqrt[3]{x} - 6x + 11\ dx$

4) $\displaystyle\int 4 - \frac{8}{x^2} + x^3\ dx$

5) $\displaystyle\int \sqrt[4]{x} + 6\sqrt{x} - \frac{7}{x^2}\ dx$

8

Now we begin to encounter integrals that can be expanded. I'm sure you're familiar with *u-sub* but we are going to try and avoid that in the meantime. We'll take the scenic route and try expanding the binomial and then applying the power rule.

Should you need a refresher on expanding binomials, here ya go:

$$\left(a+b\right)^2 = a^2 + 2ab + b^2$$

$$\left(a+b\right)^3 = a^3 + 3a^2b + 3ab^2 + b^3$$

$$\vdots$$

$$\left(a+b\right)^n = \sum_{k=0}^{n} {}_nC_k \, a^{n-k} b^k$$

Try using these rules on this next integral and maybe on Prob. 9

6) $\displaystyle\int 5x^{11} - 3(x+1)^2 \, dx$

7) $\displaystyle\int 6\sqrt[7]{x} - \frac{8}{x^2} + 8 \, dx$

8) $\displaystyle\int 6x^2 - 4x + 1 - \sqrt[11]{x} \, dx$

9) $\displaystyle\int (x^2 + 3)^3 \, dx$

So what happens when we begin to deal with integration problems involving fractions?

Well, first, start with the basics:

Is it being divided by a single term? If the answer is *yes* then, shoot, we're off the hook! Divide each term on the numerator by the denominator we were just examining. Allow me to show you:

$$\int \frac{3x^2 - 6x + 10}{\sqrt{x}} \, dx$$

$$= \int \frac{3x^2}{x^{1/2}} \, dx - \int \frac{6x}{x^{1/2}} \, dx + \int \frac{10}{x^{1/2}} \, dx$$

$$= \int 3x^{3/2} \, dx - \int 6x^{1/2} \, dx + \int 10x^{1/2} \, dx$$

... power rule time!

Notice how I not only separated these into 3 different integrals, but I also changed the radical into a rational exponent and then combined my powers. Then we were left with a power rule situation! Try it for yourself:

$$10) \int \frac{6x^3 - 4x^2 + 100}{\sqrt{x}} \, dx$$

10

I pray for your forgiveness. I know that I did not leave you with enough space to even *start* the previous problem. I do hope you have your favorite math journal or notebook next to you and are beginning to create your own canvas of math scribbles to show off...

Let's get a little crazy with the next few problems. I have not only given you some radicals, but with a binomial expansion mixed in. Be sure to watch your fractions.

11) $\int \left(4\sqrt{x} - \sqrt[3]{x}\right)^2 dx$

12) $\int x\left(\sqrt[4]{x} - 6x\right) dx$

and now we have expansions on distributions - can we *be* having any more fun?

13) $\int \left[(x^2 + 1)^2 + (x + 1)\sqrt{x}\right] dx$

14) $\int (x^4 + 1)^3 dx$

15) $\int \left[6x\sqrt[3]{x} - \frac{\sqrt{x}}{x^3}\right] dx$

16) $\int \left[15x^3 - 7(x^2 + x - 1) \right] dx$

17) $\int \left[7\sqrt{x} + 13\sqrt[4]{x} - \dfrac{1}{x^2} \right] dx$

18) $\int \left(x^3 \sqrt{x} + 2 \right)^2 dx$

19) $\int \left[4x^5 + 6x(\sqrt{x} - \sqrt[4]{x} + \dfrac{2}{x^4}) \right] dx$

Hope you felt your algebra skills tested. Unsure if you got the right answer? Check the answer key - or don't - this is truly all about learning. This is a perfect opportunity to say that the answer key in the back will look not unison. Answers will vary in format, and that's because other math nerds have solved them!

Behind the scenes, 4 math nerds have connected with me on discord and have solved these problems to give the answer key you will discover in the back of this book. I am incredibly proud of them and appreciative them, not only for their time, but for their resilience to complete each integral.

12

Trig, exponential and logarithmic integrals

The following integrals are simply all about memorizing. Yes we will have to manipulate some of the integrals presented, but let's not forget the BASICS. Guard these integral rules with you life!

Trig Integrals

$$\int \cos(u)\,du = \sin(u) + C \qquad \int \sin(u)\,du = -\cos(u) + C$$

$$\int \sec^2(u)\,du = \tan(u) + C \qquad \int \sec(u)\tan(u)\,du = \sec(u) + C$$

$$\int \csc(u)\cot(u)\,du = -\csc(u) + C \qquad \int \csc^2(u)\,du = -\cot(u) + C$$

Exponential / Log Integrals

$$\int e^u\,du = e^u + C \qquad \int \frac{1}{u}\,du = \ln|u| + C$$

$$\int a^u\,du = \frac{a^u}{\ln(a)} + C$$

I am not going to lie to you. I still forget some of these basic integrals that I reference them from time-to-time. So say it with me: *page 12, page 12, page 12.*

20) $\displaystyle\int [\sin(x) + \cos(x)]\, dx$

Of course we begin with easy ones, but don't forget that there are certain trig identities that become very helpful for these types of integrals.

21) $\displaystyle\int \sec(x)\big(\sec(x) + \cos(x)\big)\, dx$

22) $\displaystyle\int 6^x + e^x\, dx$

23) $\displaystyle\int \big(6\cos(x) + 5\sqrt{x}\big)\, dx$

24) $\displaystyle\int \left(e^x + \frac{6}{x} - \cos(x)\right) dx$

25) $\displaystyle\int 4x\left(\frac{2}{x} + \frac{\sec^2(x)}{x}\right) dx$

26) $\displaystyle\int 7^x + 7e^x\, dx$

14

The following problem *cannot* be solved using u-sub, as much as our initial instincts tell us. You should revisit some trusted identities, which (in the words of Fermat) these margins are too small to contain them. So, look them up!

27) $\int \dfrac{6\sin(2x)}{\cos(x)} + \dfrac{1}{\sqrt{x}}\,dx$

28) $\int \csc^2(x) + \sec(x)\tan(x)\,dx$

You know, these problems are so fun to write. I'm always thinking, "what would the reader expect next?" and I'm willing to bet you were not expecting another expansion-type problem. DM me on Insta if I was right!

29) $\int 2^x + 5x\left(\dfrac{1}{x} + 1\right)^2\,dx$

and how about some distribution of trig functions?

30) $\int \dfrac{1}{1 - \sin(x)}\left(\dfrac{\sin(x)}{1 + \sin(x)}\right)\,dx$

I don't quite remember where I read the following quote. It must have been on a reddit forum where OP struggled with calculus, specifically integration. Someone responded with, "you can solve any derivative out there, but can solve very little integrals."

Why is this so true? I'm not saying this because the integrals in this book are impossible, but sometimes integrals take you on an unexpected journey. This is what makes them so fascinating.

For example, we're currently covering trig and exponential integrals, thinking it'll be straight forward. However, you get something like this:

$$31) \int \sqrt{e^{2x} - 6e^x + 9} \, dx$$

and we are back to exploring what method to use. It's all about being clever, realizing that this is a quadratic in disguise that can be factored, and *hopefully* have a solution.

Give it a go and we'll go back to some familiar integrals.

$$32) \int \frac{3}{x} + e^x + 4(3)^x \, dx$$

16

33) $\displaystyle\int \frac{6x^2 + 12x + 6}{x(x+1)^2} \, dx$

Above we have another doozy. Given the skills that we have covered so far, this problem is close to unsolvable. But what happens when you play with it? (not too much, though!)

Try messing around with factoring and see what happens.

When you're finished with that, the nexts one will require some playing of their own. The fun never stops!

34) $\displaystyle\int \frac{\cos(2x)}{(\sin(x)\cos(x))^2} \, dx$

Wait, what's this?? Some helpful formulas to help guide us through these integrals. Where were they when we needed them previously?

This is life.

$$\underline{\text{Double Angle Formulas}}$$

① $\sin(2x) = 2\sin(x)\cos(x)$

② $\cos(2x) = \cos^2(x) - \sin^2(x)$

$$= 1 - 2\sin^2(x)$$

$$= 2\cos^2 x - 1$$

③ $\tan(2x) = \dfrac{2\tan(x)}{1 - \tan^2(x)}$

35) $\int e^x + \left(\dfrac{1}{x} + 1\right)^2 dx$

36) $\int 8^x - \csc^2(x) + \dfrac{\tan(x)}{\frac{1}{2}\sin(2x)} \, dx$

37) $\int \dfrac{1}{4-x} + \sec(x)\tan(x) - \sin(x) \, dx$

38) $\int \dfrac{4}{x} - \dfrac{4}{x^2} + 4^x \, dx$

39) $\int (\tan(x) + 1)^2 - \dfrac{2\sec(x)}{\csc(x)} \, dx$

Hope you're enjoying yourself. Almost 40 integrals in and I feel like we've learned so much.

Let's prep for the next adventure.

18

u-sub integration

Now that we covered the basic integrals we may encounter (along with some algebraic fun) let's review what many will agree is the most fundamental technique of integration. It's easy to master and can save you from a whole lot of trouble while out on trail.

u-sub comes in many forms and finding what to make your u-value is part of the fun! Here's an example

$$\int (2x+1)^6 \, dx \qquad u = 2x+1$$

$$du = 2\,dx$$

$$\frac{du}{2} = dx$$

* Here we find our u, our du, and isolate our dx

$$\int (u)^6 \cdot \frac{du}{2} = \frac{1}{2} \int u^6 \, du$$

Look how easy our integral looks now!

How about $\int \frac{\sin(\ln(x))}{x} \, dx \qquad u = \ln x$

$$du = \frac{1}{x} \, dx$$

$$x \, du = dx$$

Same process and it turns to

$$\int \frac{\sin(u)}{\cancel{x}} \cdot \cancel{x} \, du = \int \sin(u) \, du$$

$\cancel{x}$ be gone!

the point of u-sub integration is to decipher a function that was differentiated by chain rule. It turns our otherwise "difficult" looking integral into a basic one we recognize from previous examples.

Now, quickly identifying the u-value takes time, especially for a first-timer, but you'll eventually get the hang of it and even do alot of the work in your head.

Trust me when I saw that u-sub will show up in almost all of the problems moving forward. Sorry - we can't avoid it!

Let's get to practicing, ya nerd.

40) $\displaystyle\int (3x + 8)^5 \, dx$

Often times students ask about the constant hanging out in the integral (look at problem 41) and if we can simply pull it outside the integral. The answer is *100% yes you can!*

41) $\displaystyle\int \frac{4}{x + 8} \, dx$

42) $\displaystyle\int x(3x^2 + 1)^4 \, dx$

20

Okay, we're gonna get some good practice with the last three problems.

things are going to start heating up...just a bit.

43) $\int \left(\sin(2x) + e^{6x-1} \right) dx$

44) $\int \dfrac{3x^2 + 6}{x^3 + 6x + 8} dx$

45) $\int \left(xe^{x^2+8} - \dfrac{2}{x+3} \right) dx$

you know the next problem is a juicer when the font is *smaller*

46) $\int \dfrac{2\sin(2x)}{\cos(2x)} dx - \int \dfrac{\cos^2(x) - \sin^2(x)}{\sin(2x)} dx$

use those handy-dandy trig identities! (page 16 btw)

47) $\int x\sqrt{6x^2 - 1} \, dx$

48) $\displaystyle\int \frac{4x + 12}{x^2 + 6x - 1}\, dx$

well, I didn't think I'd show this one, but now is a good time to discuss the integral above.

Yes, it is a u-sub, where the u = *denominator* (sorry I gave it away). However, the top will not fully simplify...or will it? Try it out and you should see that we can factor out a constant to make it *2(2x+6)* and now we got things rolling.

Alright, let's continue

49) $\displaystyle\int 2\sec(2x)\tan(2x)\, dx$

50) $\displaystyle\int \frac{\sec^2(x) - \sin^2(x)}{\tan(x)}\, dx$

the next one requires some properties of exponents!

51) $\displaystyle\int e^{x^2} e^x e^2 (2x + 1)\, dx$

and didn't we see something similar to Prob 52 in Prob 31???

52) $\displaystyle\int \sqrt{e^{4x} + 2e^{2x} + 1}\, dx$

53) $\displaystyle\int \left(4^{\sin x} \cos x - \frac{3}{4x+5} \right) dx$

54) $\displaystyle\int \left(\frac{1}{\sqrt{x}(\sqrt{x}+1)} + \frac{1}{\sqrt{x}} \right) dx$

The problem above reminds of similar integrals students of mine encounter in class. At this point their instincts tell them to distribute the square root. If you try that - BAM! - you'll hit a wall.

So, try u-sub!

55) $\displaystyle\int \frac{\tan \sqrt{x}}{6\sqrt{x}} \, dx$

and for this one! 1 or 2 substitutions might be needed. It all depends on how to you want to approach it. One thing I will say is:

$$\tan(x) = \frac{\sin(x)}{\cos(x)}$$

56) $\displaystyle\int 4x(x^2+1)^5\,dx$

57) $\displaystyle\int\left(4^{5x-6}+\frac{e^x}{(e^x+1)^7}+\frac{3}{4x+1}\right)dx$

Oh, look at this! plenty of space for you to write out your work. Only if want to, of course.

24

manipulating the root

Sometimes we encounter an opportunity to not only make the value *inside* the root our u, but we invite the entire root to become our substitution. This is helps us "get rid of" the root and pay it a visit once our integral is re-written. Here's what I mean:

$$\int x^3 \sqrt{x^2+3} \; dx$$

In this example we let $u = \sqrt{x^2+3}$, and here's how we get rid of it:

$$u = \sqrt{x^2+3}$$

$$\boxed{u^2 = x^2+3} \quad \text{Now we find } dx$$

$$2u\,du = 2x\,dx$$

$$\frac{2u\,du}{2x} = dx$$

Let's rewrite the integral

$$\int x^3 u \, \frac{2u\,du}{2x} = \int x^2 u^2 \, du$$

But we know $u^2 = x^2+3$, so $x^2 = u^2 - 3$

$$\int (u^2-3)u^2\,du = \int (u^4 - 3u^2)\,du = \frac{u^5}{5} - u^3 + c$$

and at the end: $\left| \dfrac{\left(\sqrt{x^2+3}\right)^5}{5} - \left(\sqrt{x^2+3}\right)^3 + c \right.$

Some readers will apply this differently.

They would:
- make the value inside the root become the u
- find du and dx
- manipulate the u-value to accommodate the strangling x's
- solve

Here is a walkthrough using the same integral

$$\int x^3 \sqrt{x^2+3} \; dx$$

$$u = x^2 + 3$$
$$du = 2x \, dx$$
$$\frac{du}{2x} = dx$$

$$\int x^3 \cdot \sqrt{u} \; \frac{du}{2x} = \frac{1}{2}\int x^2 \sqrt{u} \; du$$

Since $u = x^2 + 3 \rightarrow u - 3 = x^2$

$$\frac{1}{2}\int (u-3)\sqrt{u} \; du = \frac{1}{2}\int (u-3)u^{1/2} \, du$$

$$= \frac{1}{2}\int u^{3/2} - 3u^{1/2} \, du$$

$$= \frac{1}{2}\left(\frac{2u^{5/2}}{5} - 2u^{3/2}\right) + C$$

$$= \frac{u^{5/2}}{5} - u^{3/2} + C$$

$$\boxed{= \frac{(x^2+3)^{5/2}}{5} - (x^2+3)^{3/2} + C}$$

26

Time to do some exploring of your own!

58) $\int 4x^5 \sqrt{x^3 + 1}\, dx$

59) $\int x^3 \sqrt{x^2 - 1}\, dx$

60) $\int 6x^3 \sqrt[3]{x^2 + 1}\, dx$

I heard you like roots, so I gave you a root multiplied to a root within another root!

61) $\int \sqrt{x}\sqrt{\sqrt{x} + 6}\, dx$

62) $\int \dfrac{x^3}{\sqrt{x^2 + 1}}\, dx$

The next one is kinda fun! It starts the same way, but you'll find yourself manipulating the u-value just a *little bit* differently this time.

63) $\int \dfrac{x^5}{\sqrt{x^2 - 1}}\, dx$

64) $\int x^5 \sqrt{x^2 + 10}, dx$

please ignore that comma - I don't know how it got there. Well, I do - it was a typo.

If you didn't already know I am the self-proclaimed "King of Typos". I try my best to re-read my problems numerous times, but there is always some typo that other math nerds point out to me.

I am human and mistakes are meant to be embraced. I have no shame in showing my cracks as they can be sealed.

That being said, should you find any typos in my work (in this book or other published work) *please* point them out to me and I will be sure to make the necessary adjustments and be forever grateful to your services.

65) $\int 10x^3 \sqrt[3]{x^2 + 8} \, dx$

66) $\int \dfrac{6x^3 - x}{\sqrt{x^2 - 1}} \, dx$

67) $\int (x^5 + x^3)(x^2 + 2)^{1/2} \, dx$

28

special trig integrals

I have done countless posts on Instagram, filmed tons of videos for TikTok/YouTube, and have performed lots of Integration Practice LIVES and you can almost guarantee the most technical of integrals always include forms that I am about to show you.

A) $\int \dfrac{1}{u^2 + a^2}\, du = \dfrac{1}{a} \arctan\left(\dfrac{u}{a}\right) + c$

B) $\int \dfrac{1}{u^2 - a^2}\, du = \dfrac{1}{2a} \ln\left|\dfrac{u-a}{u+a}\right| + c$

c) $\int \sec^3(u)\, du = \dfrac{1}{2}\sec(u)\tan(u) + \dfrac{1}{2}\ln\left|\sec(u) + \tan(u)\right| + C$

D) $\int \sec(u)\, du = \ln\left|\sec(u) + \tan(u)\right| + C$

E) $\int \cos^2(u)\, du = \dfrac{1}{2}u + \dfrac{1}{4}\sin(2u) + C$

F) $\int \sin^2(u)\, du = \dfrac{1}{2}u - \dfrac{1}{4}\sin(2u) + C$

The proofs for integrals A, B, and C will be shown later in the book since they need another skillset I've yet to cover. But, I will show you D, E, and F right now.

D) $\displaystyle\int \sec(u)\,du \frac{\big(\sec(u)+\tan(u)\big)}{\big(\sec(u)+\tan(u)\big)}$

$= \displaystyle\int \frac{\sec^2(u)+\sec(u)\tan(u)}{\sec(u)+\tan(u)}\,du$

Make $w = \sec(u)+\tan(u)$

$\qquad dw = \sec(u)\tan(u)+\sec^2(u)\,du$

$= \displaystyle\int \frac{dw}{w} = \ln|w| + C = \boxed{\ln\big|\sec(u)+\tan(u)\big| + C}$

the proof to this one is soooo simple it's almost a crime for it to be straight forward.

The proofs for E and F are very similar in technique so I will go ahead and show you E in its entirety and leave you with a teaser for F. Also, I will not be like Fermat and say the margin is too small to contain it...instead I will include the proof on the next page.

troll alert

E) $\int \cos^2(u)\, du$

Use the half angle formula

$$\cos^2(u) = \frac{1 + \cos(2u)}{2} = \frac{1}{2} + \frac{\cos(2u)}{2}$$

$$\int \cos^2(u)\, du = \int \frac{1}{2} + \frac{\cos(2u)}{2}\, du$$

$$\boxed{= \frac{1}{2}u + \frac{1}{4}\sin(2u) + C}$$

F) $\int \sin^2(u)\, du$

Use half angle formula Again

$$\sin^2(u) = \frac{1 - \cos(2u)}{2} = \frac{1}{2} - \frac{\cos(2u)}{2}$$

and the same steps as (E) Follows!

Hope that helps. Again, as for A, B, and C we will prove those later. In the meantime, keep these shortcut integrals in mind for the next few problems we have.

Wait, I wound't even suggest to keep them in mind - MEMORIZE THEM!! You'll thank me later.

68) $\displaystyle\int \frac{1}{(x+2)^2 + 25}\,dx$

Let's start by using a u-sub here (hint: x+2) and making a = 5 then see if you recognize the integral mentioned in the beginning of this chapter.

Try the same method for the next problem.

69) $\displaystyle\int \frac{1}{(3x+1)^2 - 16}\,dx$

And for this next one, we will continue using our trig identities.

70) $\displaystyle\int \frac{1 - \sin(x)}{\cos(x)}\,dx$

and just like learning to ride a bike, the training wheels are removed and off you go!

71) $\displaystyle\int \sec^3(4x)\,dx$

72) $\displaystyle\int \sin^2(3x+1) + \cos^2(4x+1)\,dx$

73) $\displaystyle\int \frac{3x}{(x^2+1)^2 + 36}\,dx$

74) $\displaystyle\int \cos^2(3x + \frac{\pi}{2})\, dx$

This next one is where your radical skills are put to the test. Let's see what you'll make your a-value equal to!

75) $\displaystyle\int \frac{1}{x^2 - 13}\, dx$

76) $\displaystyle\int \frac{6}{(4x + 5)^2 + 10}\, dx$

77) $\displaystyle\int \frac{1}{\sec(x) - \tan(x)\sin(x)}\, dx$

78) $\displaystyle\int \frac{\sec(\theta)\sin(\theta)}{\tan(\theta) + \cot(\theta)}\, d\theta$

79) $\displaystyle\int \frac{1 - \tan^2(x)}{1 + \tan^2(x)}\, dx$

completing the square review

Boo!!!! Seriously, why do I even have this section? If you have followed me for some time you'll know that I just *hate* completing the square, for no reason other than the way it's taught.

As a private math tutor I have met with hundreds of students from schools all across the country and let me tell you this: completing the square is either taught differently, or not taught at all (what???) Another reason why American academics is suffering.

But nonetheless let's do a brief review of the topic. We will start with a non-factorable quadratic with the leading coefficient a = 1

$$x^2 + bx + c \qquad\qquad x^2 + 8x - 1$$

$$\left(\frac{b}{2}\right)^2 = \frac{b^2}{4} \qquad\qquad \left(\frac{8}{2}\right)^2 = 16$$

$$x^2 + bx + \frac{b^2}{4} + c - \frac{b^2}{4} \qquad x^2 + 8x + 16 - 1 - 16$$

$$\left(x + \frac{b}{2}\right)^2 + c - \frac{b^2}{4} \qquad \left(x + 4\right)^2 - 17$$

The point is to take your b-value (the one multiplying *x*) and dividing it by 2, then squaring it. As we introduce this to our equation we need to make sure to subtract it at the end. This makes sure that we preserve our equation. From there we factor into a perfect square with some constants chilling at the end.

34

I also gave you an example to see it play out irl.

When the a-value is NOT equal to 1 the situation gets a little tricky:

$$ax^2 + bx + c$$

$$a\left(x^2 + \frac{b}{a}x\right) + c$$

$$\left(\frac{b}{2a}\right)^2 = \frac{b^2}{4a^2}$$

$$a\left(x^2 + \frac{b}{a}x + \frac{b^2}{4a^2}\right) + c - \frac{b^2}{4a}$$

$$a\left(x + \frac{b}{2a}\right)^2 + c - \frac{b^2}{4a}$$

$$3x^2 + 12x + 21$$

$$3\left(x^2 + 4x\right) + 21$$

$$\left(\frac{4}{2}\right)^2 = 4$$

$$3\left(x^2 + 4x + 4\right) + 21 - 12$$

$$3\left(x + 2\right)^2 + 9$$

In this scenario we first need to factor out a from the first two terms. This gives us a factored form with a "new" b-value. The same process is applied: divide it by 2 and square it. But watch those fractions.

We now introduce our new value inside the parantheses, but check out what happens on the outside. Instead of subtracting by the value we found, we'll need to first multiply it by a. Since the term is inside the parentheses we need recognize the a-value and subtract by the actual value we just introduced. I left an example for you to look at.

Try simplifying it and see if you get the same expression as the original.

completing the square gives TWO special integrals

I'd leave some exercises for you to complete the square, but trust me when I say we'll get plenty of practice in this section.

Check out the integrals below. The denominators have non-factorable quadratics, but if we complete the square on both of them you'll discover that some familiar integrals pop up. They are the ones I talked about in Page 28 - yay!

So, *if an integral has a non-factorable quadratic in its denominator, completing the square saves the day!*

$$\int \frac{1}{x^2+8x-1} \, dx = \int \frac{1}{(x+4)^2 - 17} \, dx$$

$$\text{Use } \int \frac{1}{u^2-a^2} \, du = \frac{1}{2a} \ln \left| \frac{u-a}{u+a} \right| + c$$

$$\int \frac{1}{3x^2+12x-1} \, dx = \int \frac{1}{3(x+2)^2+9} \, dx$$

$$\text{Factor out a } \tfrac{1}{3}: \quad \frac{1}{3}\int \frac{1}{(x+2)^2+3} \, dx$$

$$\text{Use } \int \frac{1}{u^2+a^2} \, du = \frac{1}{a} \arctan\left(\frac{u}{a}\right) + c$$

36

Ready to take a crack at some of these? Let's goooo

80) $\displaystyle\int \frac{1}{x^2 + 10x + 61}\, dx$

81) $\displaystyle\int \frac{1}{x^2 + 10x + 1}\, dx$

82) $\displaystyle\int \frac{1}{2x^2 + 12x + 11}\, dx$

83) $\displaystyle\int \frac{1}{3x^2 + 12x + 80}\, dx$

84) $\displaystyle\int \frac{e^x}{e^{2x} + e^x + 6}\, dx$

85) $\displaystyle\int \frac{\cos x}{4 \sin^2 x + 10 \sin x - 3}\, dx$

86) $\displaystyle\int \frac{1}{3x^2 - 5x + 10}\, dx$

I threw a little curveballs there, but I certainly hoped you enjoyed working them out.

One of the reasons why I love integrals so much is how these little beauties put your algebra skills to the test. One small mistake, or something gets over-looked, and you're toast!

The more you practice these methods, the better you're skillset will get. These skills can be transferred to your career in STEM, or future math classes. One day, you'll remember these gentle giants and thank them for their service.

38
Trig Substitution: The Battle of Triangles

These were taught to me in high school (shoutout Mr. Bolen!!!) as a series of formulas we had to remember. It wasn't until I sat down and drew my right triangles that I found the beauty injected in these type of integrals. They come in many forms, but here are some of the most recognizable:

A) $$\int \sqrt{u^2 - a^2}\, du \quad , \quad \int \frac{1}{\sqrt{u^2 - a^2}}\, du$$

B) $$\int \sqrt{a^2 - u^2}\, du \quad , \quad \int \frac{1}{\sqrt{a^2 - u^2}}\, du$$

C) $$\int \sqrt{u^2 + a^2}\, du \quad , \quad \int \frac{1}{\sqrt{u^2 + a^2}}\, du \quad , \quad \int \frac{1}{u^2 + a^2}\, du$$

The clue that trig-sub must be used here is the inability to do *u-substitution*. It won't work here!

So, instead of giving up, crumbling our paper, throwing it on the ground, lighting it on fire, and find a corner to cry in, we are given a pretty convenient guide to follow.

WOW!

A) set $a\sec\theta = u$, $a\tan\theta = \sqrt{u^2 - a^2}$

B) set $a\sin\theta = u$, $a\cos\theta = \sqrt{a^2 - u^2}$

c) set $a\tan\theta = u$, $a\sec\theta = \sqrt{u^2 + a^2}$

This is great to have handy and memorize. Some students prefer this, like most of my classmates in high school. But since I'm a libra and in touch with feelings I like for these integrals to have a bit more meaning. So we'll paint them a beautiful canvas!

The next three pages will include the right triangle set up for every specific integral and what they simplify to. A few things to note:

1) The integral we reviewed together on Page 28 will be proven in part C
2) The integrals are not fully solved. It will be your job to rewrite them in terms of u (or x if that's the variable you're using)

Enjoy the proofs!

Part A

$$\int \sqrt{u^2 - a^2} \, du \, , \quad \int \frac{1}{\sqrt{u^2 - a^2}} \, du$$

Let's draw the right triangle

$\leftarrow \sqrt{u^2 - a^2}$ by pythagorean theorem

Using SOH CAH TOA

$$a \sec \theta = u \quad \text{and} \quad a \tan \theta = \sqrt{u^2 - a^2}$$

$$a \sec \theta \tan \theta \, d\theta = du$$

So,
$$\int \sqrt{u^2 - a^2} \, du = \int a \tan \theta \cdot a \sec \theta \, d\theta$$

$$= \int a^2 \tan^2 \theta \sec \theta \, d\theta$$

$$\int \frac{1}{\sqrt{u^2 - a^2}} \, du = \int \frac{1}{a \tan \theta} \cdot a \sec \theta \tan \theta \, d\theta$$

Look this one up!

$$\longrightarrow \quad = \int \sec \theta \, d\theta$$

Part B

$$\int \sqrt{a^2 - u^2} \, du \, , \quad \int \frac{1}{\sqrt{a^2 - u^2}} \, du$$

$\sqrt{a^2 - u^2}$ by pythagorean Theorem

Using SOH CAH TOA

$$a \sin \theta = u \quad , \quad a \cos \theta = \sqrt{a^2 - u^2}$$

$$a \cos \theta \, d\theta = du$$

So, $\int \sqrt{a^2 - u^2} \, du = \int a \cos \theta \cdot a \cos \theta \, d\theta$

$$= \int a^2 \cos^2 \theta \, d\theta$$

$$\int \frac{1}{\sqrt{a^2 - u^2}} \, du = \int \frac{1}{a \cos \theta} \cdot a \cos \theta \, d\theta$$

$$= \int 1 \, d\theta$$

$$\int \sqrt{u^2 + a^2}\, du \quad, \quad \int \frac{1}{\sqrt{u^2 + a^2}}\, du \quad, \quad \int \frac{1}{u^2 + a^2}\, du$$

$$a\tan\theta = u$$

$$a\sec^2\theta\, d\theta = du$$

$$a\sec\theta = \sqrt{u^2 + a^2}$$

$$a^2\sec^2\theta = u^2 + a^2$$

So,
$$\int \sqrt{u^2 + a^2}\, du = \int a\sec\theta \, a\sec^2\theta \, d\theta$$

$$= \int a^2 \sec^3\theta \, d\theta$$

$$\int \frac{1}{\sqrt{u^2 + a^2}}\, du = \int \frac{1}{a\sec\theta}\, a\sec^2\theta \, d\theta$$

$$= \int \sec\theta \, d\theta$$

$$\int \frac{1}{u^2 + a^2}\, du = \int \frac{1}{a^2\sec^2\theta} \cdot a\sec^2\theta \, d\theta$$

$$= \int \frac{1}{a}\, d\theta$$

Finish up this integral, rewrite in terms of u and see how this is integral A from page 28

There you have it. I think it's finally time to do some problems!

Let's start with some basics:

87) $\displaystyle\int \sqrt{x^2 - 16}\, dx$

88) $\displaystyle\int \sqrt{x^2 + 49}\, dx$

89) $\displaystyle\int \sqrt{25 - x^2}\, dx$

90) $\displaystyle\int \frac{1}{\sqrt{4 - x^2}}\, dx$

91) $\displaystyle\int \frac{1}{\sqrt{x^2 + 10}}\, dx$

92) $\displaystyle\int \frac{1}{\sqrt{x^2 - 81}}\, dx$

Okay, I hope you got some practice under your belt, cause we're about to get crazyyyy.

44

Remember ye olde completing the square? Well, it's back to haunt us! Check out what happens when we not only have to complete the square within the square root, but also use trig sub.

93) $\displaystyle\int \frac{1}{\sqrt{x^2 + 6x - 72}}\, dx$

94) $\displaystyle\int \sqrt{x^2 - 6x + 13}\, dx$

95) $\displaystyle\int \sqrt{-x^2 + 2x + 8}\, dx$

96) $\displaystyle\int \frac{1}{\sqrt{9x^2 + 6x - 4}}\, dx$

What did you think? More of this later on (trust me, it won't go away unfortunately) but let's solve some tame-looking integrals.

97) $\displaystyle\int \frac{\cos(x)}{\sqrt{1 + \sin^2(x)}}\, dx$

98) $\int \dfrac{1}{x^2\sqrt{9x^2 - 25}}\, dx$

99) $\int \dfrac{1)}{x^2 + 2x + 5}\, dx$

We've made it to 100

100) $\int \dfrac{x}{\sqrt{16 - x^4}}\, dx$

101) $\int \dfrac{1}{\sqrt{x^2 + 6x - 16}}\, dx$

Watch your negatives on the next one.

102) $\int \dfrac{1}{\sqrt{3 - 2x - x^2}}\, dx$

103) $\int \dfrac{\sqrt{1 + x^2}}{x}\, dx$

104) $\displaystyle\int \frac{x}{\sqrt{x^2 + x + 1}}\, dx$

105) $\displaystyle\int \frac{1}{\sqrt{x^2 - 6x + 13}}\, dx$

106) $\displaystyle\int \frac{5x}{\sqrt{2x^4 + 3}}\, dx$

107) $\displaystyle\int \frac{4}{x\sqrt{7 - (x + 1)^2}}\, dx$

I'm not gonna lie, these can be a pain in the ass sometimes. They are long in terms of finding the correct u-sub AND trig sub, solving the integral, then rewriting in terms of x.

They are so important to recognize, though. They will continue to show up time and time again, especially in the next type of integrals I am about to introduce.

Seriously, take a break from integration if you want to, cause the next few pages are going to be exhausting.

Partial Fraction Decomposition

So far we have had integrals of this form

$$\int \frac{1}{ax^2 + bx + c}\, dx$$

but they have been able to be solved using completing the square, then recognizing them as integrals discussed on page 28. Completing the square was our method of choice since the quadratic was not factorable.

Well, imagine if they were in fact factorable? Now what?! This is where Partial Fraction Decomposition (PFD) comes into play. It's an exciting trick, but can become very annoying.

Consider the following integral. It's perfectly factorable on the bottom, so let's rewrite it:

$$\int \frac{2}{x^2 + 8x + 15}\, dx \implies \int \frac{2}{(x+3)(x+5)}\, dx$$

Nice. The next step begins the PFD process. We'll focus on just the value and ignore the integral for now. Since the denominator is a product of two factors then we can rip this expression apart into a sum of two fractions, where the denominators are the separate factors we originally found.

48

Here it is in play:

$$\frac{2}{(x+3)(x+5)} = \frac{A}{(x+3)} + \frac{B}{(x+5)}$$

and now we'll sorta work backwards and try to combine the terms on the right side. We need common denominators so we'll introduce (x+5) to the first fraction and (x+3) to the second.

$$\frac{2}{(x+3)(x+5)} = \frac{A(x+5)}{(x+3)(x+5)} + \frac{B(x+3)}{(x+3)(x+5)}$$

$$\frac{2}{(x+3)(x+5)} = \frac{A(x+5) + B(x+3)}{(x+3)(x+5)}$$

$$2 = A(x+5) + B(x+3)$$

Combining the two fractions gives an expression that is equal to the left hand side, since both denominators are equal then it leaves us no choice but to make both numerators equal to each other.

Now our job is to find the values of A and B.

We'll use something called "cover up" method. The trick is to cover up one of the parentheses by setting x equal to a value that'll turn the parentheses into zero. Check this out

$$X = -5 \rightarrow 2 = A(-5+5) + B(-5+3)$$

$$2 = -2B$$

$$-1 = B$$

$$X = -3 \rightarrow 2 = A(-3+5) + B(-3+3)$$

$$2 = 2A$$

$$1 = A$$

Doing this helps us find the values of A & B, then go back to our integral

$$\int \frac{2}{(x+3)(x+5)} \, dx = \int \frac{A}{(x+3)} \, dx + \int \frac{B}{x+5} \, dx$$

$$= \int \frac{1}{x+3} \, dx + \int \frac{-1}{x+5} \, dx$$

$$= \ln|x+3| - \ln|x+5| + C$$

Since we now know the correct values of A and B our integral becomes easy to solve using a natural log discussed earlier in this book.

Let's play a little bit.

108) $\displaystyle\int \frac{8}{x^2 + 4x - 12}\, dx$

109) $\displaystyle\int \frac{1}{x^2 + 3x}\, dx$

don't forget your GCF on this one!

110) $\displaystyle\int \frac{-x + 10}{x^2 + x - 12}\, dx$

111) $\displaystyle\int \frac{3x - 2}{x^2 + 5x + 4}\, dx$

112) $\displaystyle\int \frac{5x - 46}{x^2 - 3x - 28}\, dx$

113) $\displaystyle\int \frac{5x + 1}{x^2 - x - 2}\, dx$

114) $\displaystyle\int \frac{1}{x^3 - x}\, dx$

We're keeping it tame right now. So far, every single expression factored has been relatively easy (if you didn't think so it's a sign you need to revisit factoring quadratics. The more practice the better the skills) but what happens when you end up having a non-factorable quadratic *after* we factor out initially?

$$\int \frac{x^2 - 2x + 2}{x^3 + x^2 + 4x + 4} \, dx$$

$$\int \frac{x^2 - 2x + 2}{x^2(x+1) + 4(x+1)} \, dx = \int \frac{x^2 - 2x + 2}{(x^2 + 4)(x+1)} \, dx$$

Check this integral out. We broke the denominator down by grouping, then we're left with a quadratic as one of our factors.

The same rule will apply. We rewrite the rational expressions as a sum of two fractions, but look at how write the one with the quadratic.

$$\frac{x^2 - 2x + 2}{(x^2 + 4)(x + 1)} = \frac{A}{(x+1)} + \frac{Bx + C}{(x^2 + 4)}$$

It's important to write the number as a linear expression, since our denominator was quadratic (degree 2).

52

We continue by adding the two fractions by getting common denominators and make it equal to the left hand side.

In this particular instance, unfortunately, the "cover up" method cannot be used. Instead, we need to distribute and combine like terms and make each term equal to the term values that we have on the left.

We then solve using system of equations.

$$\frac{x^2-2x+2}{(x^2+4)(x+i)} = \frac{A}{(x+i)} + \frac{Bx+C}{(x^2+4)}$$

$$x^2-2x+2 = A(x^2+4) + (x+i)(Bx+C)$$

$$x^2-2x+2 = Ax^2+4A + Bx^2+Cx+Bx+C$$

$$x^2-2x+2 = (A+B)x^2 + (B+C)x + (4A+C)$$

(1) $A+B=1 \rightarrow B=1-A$

(2) $B+C=-2 \rightarrow (1-A)+C=-2$
$$-A+C=-3$$

(3) $4A+C=2$

$$\text{Solve:} \quad -A+C=-3 \qquad \boxed{A=1} \;\; \boxed{C=-2}$$
$$4A+C=2$$
$$\boxed{B=0}$$

Once we find the values for A, B, and C, we then start integrating using methods discussed in the past.

$$\int \frac{x^2-2x+2}{(x^2+4)(x+1)}\,dx = \int \frac{A}{(x+1)}\,dx + \int \frac{Bx+C}{(x^2+4)}\,dx$$

$$= \int \frac{1}{(x+1)}\,dx + \int \frac{-2}{(x^2+4)}\,dx$$

$$\boxed{= \ln|x+1| - \arctan\left(\frac{x}{2}\right) + C}$$

I know - whatttt?! Yeah, that's integration for you. Sometimes an innocent integral can turn more complex by switching a few things, but this is why I have outlined the book this way - we're taking it one step at a time.

Alright, time for you to attempt some of these!

115) $\int \dfrac{3x^2 - 6x + 4}{(x-3)(x^2+4)}\,dx$

116) $\int \dfrac{x^2 - 2x + 10}{(x^2+4)(x-3)}\,dx$

117) $\int \dfrac{-x^2 + 6x - 15}{x^3 - x^2 + 9x - 9}\,dx$

118) $\displaystyle\int \frac{x^2 + 2x - 1}{(x-1)(x^2+1)}\,dx$

119) $\displaystyle\int \frac{12x^2 + x + 109}{3(x+1)(x^2+9)}\,dx$

120) $\displaystyle\int \frac{4x^2 - 56}{(x^2-9)(x^2+1)}\,dx$

this next one is a classic! Try using sum/difference of cubes and go from there!

121) $\displaystyle\int \frac{1}{x^3 + 1}\,dx$

122) $\displaystyle\int \frac{2}{9x^3 - 243}\,dx$

123) $\displaystyle\int \frac{2}{x^5 - x^2}\,dx$

Will you hate me if I tell you there's *another* scenario to look out for when it comes to PFD?

Awww, that's so sweet of you! Trust me, I too am a little tired but it would be a disservice if this method was not presented.

Below we have a situation where our denominator was a perfect square. Well, we still need to write it as a sum of two fractions, and we do so by making sure to account for every power!

in this case since it was only a power of 2 we stopped there. but if it was a power of 3 we would continue and add another fraction accounting for that power.

Lastly, notice how our factor was a linear term so we only need to write an A or B as our numerator.

$$\int \frac{2x-1}{(x-1)^2} \, dx$$

$$\frac{2x-1}{(x-1)^2} = \frac{A}{(x-1)} + \frac{B}{(x-1)^2}$$

Now we go ahead and combine these fractions by getting common denominators. Oh, what do you know, the fraction with the B terms doesn't need anything, only the A term!

$$2x - 1 = A(x-1) + B$$

$$x = 1 \longrightarrow \quad 2(1) - 1 = A(1-1) + B$$

$$\boxed{1 = B}$$

$$x = 2 \quad 2(2) - 1 = A(2-1) + 1$$

$$3 = A + 1$$

$$\boxed{2 = A}$$

$$\int \frac{2x-1}{(x-1)^2}\, dx = \int \frac{2}{x-1}\, dx + \int \frac{1}{(x-1)^2}\, dx$$

$$\boxed{= 2\ln|x-1| - \frac{1}{x-1} + c}$$

We end up following the same method here and find our solution to the integral. Yay, PFD!

Ready for some problems of your own?

124) $\displaystyle\int \frac{2x - 2}{x^2 + 4x + 4}\, dx$

125) $\displaystyle\int \frac{x^2 - 2x - 1}{(x - 1)^2(x^2 + 1)}\, dx$

126) $\displaystyle\int \frac{x^2}{x^3 - 3x^2 + 3x - 1}\, dx$

127) $\displaystyle\int \frac{4x^3 + 8x^2 - 10x - 10}{(x + 1)^2(x^2 - 5)}\, dx$

128) $\displaystyle\int \frac{-4x^2 - 21x - 385}{(x + 5)(x^2 + 6x + 100)}\, dx$

129) $\displaystyle\int \frac{8x^3 + 4x^2 + 71x + 20}{(x + 1)^2(x^2 + 10)}\, dx$

130) $\displaystyle\int \frac{x^3 - 2x - 3}{x^4 + 4x^3 + 6x^2 + 4x + 1}\, dx$

I think it's time for a break. Thoughts?

58

Long Division THEN PFD

Okay, we just completed a hefty section of integration. Whether you find PFD difficult or not we can definitely agree that they are some long ass problems!

Admittedly, they are one of my favorite types of integrals. We get taken into a whole different algebraic world that truly test our skills, always expecting us to make the wrong move.

I want to emphasize that PFD will not go away. We will definitely be seeing it for sections to come. So, be prepared - just saying!

Let's explore what happens when we encounter integrals involving rational functions where **the degree of numerator is greater than or equal to the degree of denominator**

This is where Long Division comes into play.

$$\int \frac{x^2 - 7x - 11}{x - 8} \, dx \qquad \int \frac{x^2 - x - 29}{x - 6}$$

$$\int \frac{4x^3 + 6x^2 - 10x + 4}{2x - 1} \, dx \qquad \int \frac{x^2 + 6x - 1}{x^2 + 3x - 4} \, dx$$

Look at those integrals very carefully. Every single one of them has the degree of numerator greater than or equal to the degree of denominator.

And while some of them look like PDF can be used, attemping that first will lead to a dead end. (I double dare you to try it!)

So instead you must first simplify these rational functions using long or synthetic division. I encourage you to revisit long and synthetic division if you don't remember it - especially synthetic - because it will also be covered in the next section.

I'll leave you with one final note: after dividing this the rational function turns into something much simplier and recognizable. Let's see if you've got the skill set.

131) $\int \dfrac{x^2 - 7x - 11}{x - 8}\, dx$

132) $\int \dfrac{x^2 - x - 29}{x - 6}\, dx$

133) $\int \dfrac{4x^3 + 6x^2 - 10x + 4}{2x - 1}\, dx$

134) $\int \dfrac{x^2 + 6x - 1}{x^2 + 3x - 4}\, dx$

60

Hope you enjoyed the "easy" ones, but let's continue this madness.

135) $\displaystyle\int \frac{x^2 + 1}{x^2 + 3x + 2}\, dx$

136) $\displaystyle\int \frac{3x^3 - x + 12}{x^2 - 1}\, dx$

137) $\displaystyle\int \frac{2x^2 - 9x - 10}{x^2 - 5x}\, dx$

138) $\displaystyle\int \frac{x^4}{x^2 - 1}\, dx$ whoops! make this a 3

139) $\displaystyle\int \frac{x^4 - 2x - 11x^2 + 34}{x + 2}\, dx$

140) $\displaystyle\int \frac{2x^3 + 4x^2 + 5}{x - 2}\, dx$

141) $\displaystyle\int \frac{x^4 + 2x^3 - 4x^2 + 3x - 5}{x^2 + 3x - 2}\, dx$

142) $\displaystyle\int \frac{x^3 - 4x^2 + x + 7}{x + 1}\, dx$

Don't fall for the trick I am about to give you on the next one!

143) $\displaystyle\int \frac{3x^3 + 12x - x^2 - 3}{x^2 + 4}\, dx$

144) $\displaystyle\int \frac{5x^3 - 9x^2 + 24x - 5}{5x - 4}\, dx$

145) $\displaystyle\int \frac{6x^4 - 2x^3 - 3x + 2}{x^2 + 1}\, dx$

146) $\displaystyle\int \frac{3x^5 - x^3 + 4x^2 - 12x - 8}{x^2 - 2}\, dx$

147) $\displaystyle\int \frac{x^5 + x^4 - x^3 + 3}{2x - 1}\, dx$

148) $\displaystyle\int \frac{x^3 + 6x^2 + 4x - 7}{x^3 + 1}\, dx$

Wow. What just happened? I can't believe that we just covered, what I think, is the most exhausting section of integration.

there's a noise in the distance

Wha-what is that? It's coming closer. I'm scared - you feel that?

the floor violently shakes

Oh no! What's going on?? Let me call out to this entity.

Hello? Anybody there?

"You think you are finished? We're just getting started!!!"

This isn't good.

"I come before you to block anything that comes through this path. You shall not pass unless you can defeat me! It is I, the Rational Root Theorem! GRRRRRRR!"

Damn...

Rational Root Theorem

I honestly contemplated leaving this section out of this book. It's long, ugly, and almost no one likes it.

But it's just a beast of a method once in a while we have to face. First let's start with what this theorem states.

Rational Root Theorem

A rational root of polynomial function
$$f(x) = a_n x^n + a_{n-1} x^{n-1} + ... + a_2 x^2 + a_1 x + a_0$$

is of the form:
$$\frac{p}{q} = \frac{\text{factor of last term } a_0}{\text{factor of first term } a_n}$$

Okay, easier said than done, but *how* and *when* do we use it?

It all starts when we have an innocent looking integral:

$$\int \frac{6x^2 - 10x + 8}{5x^3 + 29x^2 + 19x - 5} \, dx$$

64

Shoot. Well, we can't use long division since it does not satisfy the conditions.

PFD can't be done since it can't factored...yet! This is when Rational Root Theorem is used.

Let's take a look at that theorem. The factors of a polynomial is a ratio of the factors of the constant value and the leading coefficient. Since I am hinting that PFD could be used we will focus on factoring out the denominator !

$$5x^3 + 29x^2 + 19x - 5$$

$$\frac{P}{q} = \frac{\text{Factors of the constant}}{\text{Factors of the leading coefficient}}$$

$$= \frac{\pm 1, \pm 5}{\pm 1, \pm 5}$$

$$= \frac{\pm 1}{\pm 1}, \pm \frac{1}{5}, \frac{\pm 5}{1}, \pm \frac{5}{5}$$

$$= \pm 1, \pm \frac{1}{5}, \pm 5, \pm 1$$

$$= \pm 1, \pm \frac{1}{5}, \pm 5$$

Yes, they both happen to be 5, but make sure to put the factors of the constant on top, and factors for leading coefficient on the bottom.

I went ahead and wrote all the positive and negative factors on a list and now I found every single combination of ratios while getting rid of repeating factors.

Okay, time for some synthetic division. Now that we have our *possible* roots we need to sorta guess which one is could be.

This is the sucky part - we have to hope for the best and eliminate the ones that don't work along the way.

$$
\begin{array}{r|rrrr}
 & 5 & 29 & 19 & -5 \\
1 & \downarrow & 5 & 34 & 53 \\
\hline
 & 5 & 34 & 53 & \boxed{48}
\end{array}
$$

checking if 1works...nope, it doesn't!

$X = 1$ is not a root!

$(x-1)$ is not a factor

$$
\begin{array}{r|rrrr}
 & 5 & 29 & 19 & -5 \\
-1 & \downarrow & -5 & -24 & 5 \\
\hline
 & 5 & 24 & -5 & \boxed{0}
\end{array}
$$

but -1 does work - whoo hoo!

$X = -1$ is a root!

$(x+1)$ is a factor

Now that we found one of them, we could continue using synthetic division to find the rest, but the quotient is a quadratic so we can just factor out the rest!

$$5x^3 + 29x + 19x - 5 = (x+1)(5x^2 + 24x - 5)$$
$$= (x+1)(5x - 1)(x + 5)$$

$$\int \frac{6x^2 - 10x + 8}{5x^3 + 29x^2 + 19x - 5} \, dx = \int \frac{6x^2 - 10x + 8}{(x+1)(5x-1)(x+5)} \, dx$$

Once we have the factored form our original integral can be written and off we go with PFD. Told ya it's here to stay!

Let's see if you can finish off this problem.

$$149) \int \frac{6x^2 - 10x + 8}{5x^3 + 29x^2 + 19x - 5} \, dx$$

Let's continue the fun, my friend.

$$150) \int \frac{7x^2 + 33x + 32}{x^3 + 4x^2 + x - 6} \, dx$$

$$151) \int \frac{x^2 - 12x + 10}{-2x^3 + x^2 - 8x + 4} \, dx$$

152) $\displaystyle\int \frac{5x^4 - x^3 - 24x^2 - 30x - 150}{x^5 - x^4 - 5x^3 + 5x^2 - 36x + 36}\, dx$

I'd like to apologize for putting you through this pain. The Rational Root Theorem monster made me do it, I swear!

We have a few more in this section and then we're done. The madness will end soon.

153) $\displaystyle\int \frac{x^2 + 14x - 5}{x^3 + 9x^2 - 21x - 2}\, dx$

154) $\displaystyle\int \frac{26x^3 + 51x^2 + 13x - 6}{5x^4 + 9x^3 + 3x^2 - x}\, dx$

155) $\displaystyle\int \frac{7x^2 + 14x - 1}{3x^3 + 11x^2 + 5x - 3}\, dx$

156) $\displaystyle\int \frac{6x^4 + x^3 - 19x^2 + 12}{2x^5 - 4x^4 - 7x^3 + 14x^2 + 6x - 12}\, dx$

Okay, I think that's enough of that. You will encounter more of these integrals in the wild. Until then, I think it's time to move on and get a little punny.

The P.U.N Integral

Prepare for **U**-sub in **N**umerator is something I came up with to help me memorize the next type of integral. If you're familiar with other methods, that's great! If you're not, today is your lucky day!

Let's talk about the PUN Integral.

$$\int \frac{2x + 3}{x^2 + 6x + 45} \, dx$$

In this particular situation the integral (A) cannot be solved using PFD and (B) completing the square won't simplify the answer (or will it?) and **most importantly** (C) the top is *almost* the derivative of the bottom.

Look at it. If we make

$$u = x^2 + 6x + 45$$

$$du = 2x + 6 \ dx$$

You will see that the top is almost the same value for du. All we need to do is have a +6....so that's what we will do to the top.

$$\int \frac{2x + 6 + 3 - 6}{x^2 + 6x + 45} \, dx$$

$$\int \frac{2x + 6}{x^2 + 6x + 45} \, dx - \int \frac{3}{x^2 + 6x + 45} \, dx$$

we add and subtract a 6 so that we have the desired du value and the remaining stuff at the end. This helped us rewrite the integral into two separate problems

$$\ln\left|x^2+6x+45\right| - \int \frac{3}{(x+3)^2+36} \, dx$$

$$\boxed{\ln\left|x^2+6x+45\right| - \frac{1}{2}\arctan\left(\frac{x+3}{6}\right) + C}$$

where the first was solved using u-sub and the second a completing the square + arctan combo.

Sometimes a PUN integral requires you to change the first term, too.

$$\int \frac{x-1}{x^2+6x+45} \, dx$$

$$\frac{1}{2}\int \frac{2(x-1)}{x^2+6x+45} \, dx = \frac{1}{2}\int \frac{2x-2}{x^2+6x+45}$$

in this case we begin by multiplying the top by 2 and bringing a 1/2 on the outside. We do that because our du value will have a 2x.

once that's complete a +6 and -6 is introduced to the top. We save the 2x + 6 and leave the rest at the end. This leaves us with two integrals where one is solved using u-sub and the second by completing the square + arctan. I love me a good pun integral!

$$\frac{1}{2} \int \frac{2x +6 -2 -6}{x^2 +6x +45} \, dx$$

$$\frac{1}{2} \int \frac{2x+6}{x^2+6x+45} \, dx + \frac{1}{2} \int \frac{-8}{x^2+6x+45} \, dx$$

Let's see your skill set with these next problems.

$$157) \int \frac{6x - 4}{3x^2 - 9x + 1} \, dx$$

$$158) \int \frac{x - 3}{2x^2 + 1x + 8} \, dx$$

For the next one, when life gives you fractions, you say "Damnnit to hell!", and keep solving.

$$159) \int \frac{5x + 1}{x^2 + 1x + 4} \, dx$$

160) $\int \dfrac{7x + 2}{3x^2 + 5x + 4} \, dx$

161) $\int \dfrac{3x - 3}{x^2 + 5x + 15} \, dx$

162) $\int \dfrac{6x - 1}{x^2 + 10x + 26} \, dx$

163) $\int \dfrac{2x + 5}{x^2 - 3x + 100} \, dx$

164) $\int \dfrac{4x + 5}{5x^2 - 20x + 1} \, dx$

165) $\int \dfrac{x - 7}{8x^2 + 3x - 1} \, dx$

166) $\int \dfrac{3x + 100}{6x^2 - 1x + 2} \, dx$

whoo, that was fun! Time to put on our wizard/witches hat, pull out our crystal balls and practice some voodoo magic.

Ultra Violet Voodoo

You can tell a lot about a person by what they call the next method.

"Integration By Parts": This person plays by the book, follows all the rules in society, and can be trusted to be a good friend.

"it's just u-sub": Stay far away from this person. I will not let them near my children.

"ILATE": probably a big nerd. Nice person to have around, but will probably correct your work and grammar.

"Ultra Violet Voodoo": The definition of the perfect human. This individual is down to earth and down for whatever. Make a math mistake? They won't judge you. Ask them to hang out? They probably will cancel all their plans to be with you. If you hear someone mention this phrase rest assured they are good people. That's all I gotta say about that!

Obviously I'm being a little facetious, but I like the ring to "Ultra Violet Voodoo".

$$\int u \, dv = uv - \int v \, du$$

Ultra Violet Voodoo

So when do we use this trick? Typically, if all the previous methods covered in this book fail, but really it's when you can identify a problem where

1) One function has an easy derivative
2) The other function is easily integratabtle .

Easier said than done, but allow me to give you an example.

$$\int \underbrace{(x)}_{u} \underbrace{\sin{(x)}\,dx}_{dv}$$

$$u = x \qquad\qquad v = -\cos(x)$$

$$du = dx \qquad\qquad dv = \sin x\,dx$$

Here we identify the x as our u-value. Why? Because taking the derivative yields an easier function (in this case it is dx) and our dv is easy enough we can integrate. *If you did this the other way around you'll find it to be more difficult* (don't believe it? try it!)

$$uv - \int v\,du$$

$$-x\cos(x) + \int \cos(x)\,dx$$

$$\boxed{-x\cos{(x)} + \sin{(x)} + C}$$

74

It's quite "simple" and I will give you some problems to practice in a second.

Before I do that let's answer a question students often ask:

"How do I know which one to choose as my u?"

The answer is ILATE:

I - Inverse Trig Functions
L - Logarithmic Functions
A - Algebraic
T - Trig Functions
E - Exponential Functions

This is the guideline you should follow. Whatever is present in a particular integral is our u value, in this particular order.

Okay, let's give it a try:

$$167) \int x \cos(x) \, dx$$

$$168) \int x^3 \ln(x) \, dx$$

$$169) \int \sin(x) \ln(\cos(x)) \, dx$$

170) $\displaystyle\int 3x^3 \sqrt{4 - x^2}\, dx$

heads up - the next problem will probably require two steps of integration by parts (ultra violet voodoo)

171) $\displaystyle\int x^2 \sin(x)\, dx$

172) $\displaystyle\int x^3 \sqrt{9 - x^2}\, dx$

173) $\displaystyle\int \frac{1}{x(\ln(x))^3}\, dx$

174) $\displaystyle\int 4x \cos(6x - 1)\, dx$

Let's finish it up with a labyrinth of a definite integral!

175) $\displaystyle\int_1^e x^4 \ln(x)\, dx$

76

Ultra Violet Voodoo, But With a Ghost

I hope you're enjoying these so far. Integration by parts can be extremely useful when tackling integrals that otherwise look almost impossible to solve.

I want to talk about about when happens when we encounter IBP with a ghost function.

$$176) \int \sin^{-1} x \, dx$$

Consider the integral above. There is only ONE function, so how the heck are we supposed to use IBP??

Here, make u = the only function given
and dv = dx (or the ghost in this case)

From there, see if you can finish off the integral.

$$177) \int \ln(x) \, dx$$

Heads up: just because we identify the proper u and dv doesn't mean the integral is easier to solve!

$$178) \int \tan^{-1} x \, dx$$

179) $\int \cos^{-1} x \, dx$

180) $\int \ln(x^2 + 1) \, dx$

181) $\int \ln(x^2 + 2x - 3) \, dx$

I gotta say - integrals like 180 and 181 are sorta fun to solve. If you have not tackled these already, we find the ghost and eventually find yourself doing a partial fraction decomposition, an arctan special integral, or perhaps some long division - who knows what will happen!

182) $\int \ln(3x^2 + x + 1) \, dx$

183) $\int \ln(x^2 - 1) \, dx$

184) $\int \ln(x^2 + 4) \, dx$

78

185) $\int \ln(x^2 - a^2)\, dx$, $a \in \mathbb{Z}$

186) $\int \ln(x^2 - 6x + 1)\, dx$

187) $\int \ln(x^2 - 10x + 1)\, dx$

We're still vibing with ultra violet voodoo, right? If not, feel free to take a break because we've got TWO more unique cases of it.

Ultra Violet Voodoo, But Move That SH!T to the Other Side

Let's hope we're pretty comfortable with this now, cause this next little trick feels like cheating.

$$188) \int e^x \cos(x)\, dx$$

Pretty innocent looking. Using ILATE we will cosine our u-value and e^x dx = dv

$$u = \cos(x) \qquad\qquad v = e^x$$
$$du = -\sin(x)\, dx \qquad dv = e^x\, dx$$

$$uv - \int v\, du$$

$$e^x \cos(x) + \int e^x \sin(x)\, dx$$

Okay, this is not *that* different. So we have another rendition of ultra violet voodoo doo, no big deal.

$$u = \sin(x) \qquad\qquad v = e^x$$

$$du = \cos(x)\, dx \qquad dv = e^x\, dx$$

$$e^x \cos(x) + e^x \sin(x) - \int e^x \cos(x)\, dx$$

This looks a bit weird since it seems as though we are back to the original integral - which it is! Look what I do next:

$$e^x \cos(x) + e^x \sin(x) - \int e^x \cos(x)\, dx = \int e^x \cos(x)\, dx$$

$$+ \int e^x \cos(x)\, dx \quad + \int e^x \cos(x)\, dx$$

$$e^x \cos(x) + e^x \sin(x) = 2 \int e^x \cos(x)\, dx$$

$$\frac{e^x \cos(x) + e^x \sin(x)}{2} + C = \int e^x \cos(x)\, dx$$

I end up setting our current work equal to the original integral and add that sh!t to the other side. I divide by 2 and we have our solution.

Nice.

K, your turn!

189) $\int e^{2x} \sin(x)\, dx$

190) $\int e^{2x} \cos(6x + 2)\, dx$

191) $\int \sin^3 x\, dx$

Now it's time to tackle the final integral outlined in Page 28

192) $\int \sec^3 x\, dx$

$\int \sec^3 x\, dx$

Let's break this apart!

$\int \sec^2 x \cdot \sec x\, dx$

$u = \sec x$

$du = \sec x \tan x\, dx$

$v = \tan x$

$dv = \sec^2 x\, dx$

Using $uv - \int v\, du$

$$\sec x \tan x - \int \tan x \cdot \sec x \tan x \, dx$$

Remember $1 + \tan^2 x = \sec^2 x$
$$\tan^2 x = \sec^2 x - 1$$

$$\sec x \tan x - \int \sec x \left(\sec^2 x - 1\right) dx$$

$$\sec x \tan x - \int \sec^3 x \, dx + \int \sec x \, dx$$

↰ Well what do we
have here!

$$\sec x \tan x - \int \sec^3 x \, dx + \int \sec x \, dx = \int \sec^3 x \, dx$$
$$+ \int \sec^3 x \, dx \qquad\qquad + \int \sec^3 x \, dx$$

$$\sec x \tan x + \int \sec x \, dx = 2 \int \sec^3 x \, dx$$

$$\sec x \tan x + \ln\left|\sec x + \tan x\right| + C = 2 \int \sec^3 x \, dx$$

$$\boxed{\frac{1}{2} \sec x \tan x + \frac{1}{2} \ln\left|\sec x + \tan x\right| + C = \int \sec^3 x \, dx}$$

Ultra Violet Voodoo, But with Tic Tac Toe

If you've seen the movie *Stand and Deliver* you'll recognize Mr. Escalante's "Tic-Tac-Toe" method. Most teachers, and books, will call this the tabular method, which is an extension of integration by parts.

But "Tabular Method" is so boring!!!! Tic Tac Toe sounds fun, exciting, provocative, and it gets the people going.

$$\int x^2 \sin x \, dx$$

Let's say you're given the integral above. Yes, you absolutely can give this the ultra violet voodoo treatment, but will soon find that you will need to do it again. I challenge you to try it for yourself!

Now, **when do we use Tic Tac Toe?**

Notice that

1) Integration by Parts is needed
2) One of our functions can be differentiated multiple times and eventually get to zero (more on that later)
3) The other function can easily be integrated.

This is the perfect recipe for Tic Tac Toe. Check it out

$$\int x^2 \sin(x)\, dx$$

x^2	$\sin(x)$
$2x$	$-\cos(x)$
2	$-\sin(x)$
0	$\cos(x)$

we integrate over and over again

I take the derivative until I get to zero!

it's important to note that you stop when the derivative becomes zero. Now add a + and - down the list...

$+$	x^2	$\sin(x)$
$-$	$2x$	$-\cos(x)$
$+$	2	$-\sin(x)$
$-$	0	$\cos(x)$

Our final step is to multiply and combine each teach in this right-diagonal fashion. There forms the tic-tac-toe!

"Tic" $+$ x^2 $\sin(x)$

"Tac" $-$ $2x$ $-\cos(x)$

"Toe" $+$ 2 $-\sin(x)$

$-$ 0 $\cos(x)$

$$+x^2(-\cos x) - 2x(-\sin x) + 2\cos x + C$$

$$\boxed{-x^2\cos x + 2x\sin x + 2\cos x + C}$$

There is going to be instances where arriving to zero on the derivative end takes a while. That is okay...continue this process with as many right-diagonal action necessary.

193) $\displaystyle\int 16x^3 \sec^3(x^4)\, dx$

194) $\displaystyle\int x^4\cos(2x)\, dx$

86

195) $\displaystyle\int x^4 e^{2x}\, dx$

196) $\displaystyle\int (x^2 + 1) \sin(2x)\, dx$

197) $\displaystyle\int (2x)^3 \cos(2x)\, dx$

198) $\displaystyle\int (2x + 1)^3 e^x\, dx$

199) $\displaystyle\int x^{10} \sin x\, dx$

200 $\displaystyle\int \frac{(2x)^5 \sin(2x + 1)}{4x^3}\, dx$

Power Reducing Integrals

Whoo - that last section was a bit exhausting! Honestly, even writing it took a while. but I'm ready to move on. Are you?!

Let's discuss some integrals of the form:

$$\int \sin^m(x) \cos^n(x)\, dx$$

$$\int \sec^m(x) \tan^n(x)\, dx$$

here both m and n are positive integers. The trick with these is breaking them down into combinations of trig integrals we recognize, apply u-sub, and end with the power rule.

Here's are the guidelines:

1) If the power of the sine is odd and positive, save one sine for your du value, then convert everything into cosines:

$$\int \sin^{2k+1} \cos^n x\, dx$$

$$= \int (\sin^2 x)^k \cos^n x \sin x\, dx$$

$$= \int (1 - cos^2 x)^k \cos^n x \sin x\, dx$$

2) If the power of the cosine is odd and positive, save one of the cosines for you du value, then convert everything into cosines:

$$\int \sin^m \cos^{2k+1} x \, dx$$

$$= \int \sin^m x (\cos^2 x)^k \cos x \, dx$$

$$= \int \sin^m x (1 - \sin^2 x)^k \cos x \, dx$$

3) In the event that both sine and cosine are even and positive, then convert using the half angle formulas in Page 28 and follow guidelines 1 and 2.

Check this out:

$$\int \sin^3 x \, \cos^4 x \, dx$$

since sine is odd we follow guideline 1

$$\int \sin^2 x \, \cos^4 x \, \sin x \, dx$$

$$\int (1 - \cos^2 x) \cos^4 x \, \sin x \, dx$$

$$u = \cos x$$

$$du = -\sin x \, dx$$

one we establish u-sub the rest is easy!

$$-\int (1 - u^2) u^2 \cdot du$$

201) $\int \sin^6(x)\cos^3(x)\,dx$

202) $\int \sin^5(x)\cos^2(x)\,dx$

203) $\int \sin^5(x)\cos^3(x)\,dx$

the integral above has BOTH sine and cosine as an odd power...guideline 1 or 2, the choice if yours!

204) $\int \sin^4(2x)\cos^5(x)\,dx$

205) $\int \cot^2(x)\sin^5(x)\,dx$

don't fear 205...simply convert that cotangent and take it from there.

206) $\int \sin^4(x)\,dx$

207) $\int \cos^2(2x)\,dx$

Now let's review the integrals that look like

$$\int \sec^m(x) \tan^n(x) \, dx$$

Here are the guidelines:

1) If the power of secant is even and positive, save a sec^2 for our du value and convert the remaining into tangents.

$$\int \sec^{2k} \tan^n x \, dx$$

$$= \int (\sec^2 x)^{k-1} \tan^n x \sec^2 x \, dx$$

$$= \int (1 + \tan^2 x)^{k-1} \tan^n x \sec^2 x \, dx$$

2) If the power of the tangent is off and positive, save a secant/tangent combo for our du value, convert the tangents into secants.

$$\int \sec^m \tan^{2k+1} x \, dx$$

$$= \int \sec^{m-1} x (\tan^2 x)^k \sec x \tan x \, dx$$

$$= \int \sec^{m-1} x (\sec^2 x - 1)^k \sec x \tan x \, dx$$

3) If there are no secants and the power of the tagent is even and positive, convert a tan^2 into the sec^2 - 1 and expand if necessary

$$\int \tan^n x \, dx$$

$$= \int \tan^{n-2} x (\tan^2 x) \, dx$$

$$= \int \tan^{n-2} x (\sec^2 x - 1) \, dx$$

4) If the integral has nothing but secants, use integration by parts (look at problem 192)

5) If none of these work, try converting into sines and cosines!!!

Honestly, let's hope we never have to use guideline 5.

Let's see you take a wack at some of these problems.

208) $\int \tan^3 x \sec x \, dx$

209) $\int \tan^5 x \, dx$

210) $\int \tan^5 x \sec^4 x \, dx$

211) $\int \sec^6 x \, dx$

212) $\int \sec^4 x \, dx$

Here to apologize for the last two problems before you get mad at me.

213) $\int \tan^2(4x - 1) \, dx$

214) $\int \tan^3(5x) \sec(5x) \, dx$

215) $\int \sec^2 \frac{x}{3} \tan \frac{x}{3} \, dx$

216) $\int \sec^5 x \, dx$

The Wallis Integrals

Believe it or not, Newton and Leibniz did not in fact invent calculus. They take credit for it (who takes the most is up for debate), but along with other great mathematicians before them one particular dude is worth mentioning.

This is John Wallis. He takes credit for the infinity symbol (weird flex, but okay) truth is John did alot of his work on calculus before the Newton and Leibniz.

But we're not here to argue who takes credit for introducing calculus to the world (it was Newton), but talk about the incredible Wallis Integral Formulas

94

The Wallis Integral Formulas:

1) If n is odd and greater than or equal to 3

$$\int_0^{\pi/2} \cos^n x \, dx = \left(\frac{2}{3}\right)\left(\frac{4}{5}\right)\left(\frac{6}{7}\right)\ldots\left(\frac{n-1}{n}\right)$$

2) If n is even and greater than or equal to 2

$$\int_0^{\pi/2} \cos^n x \, dx = \left(\frac{1}{2}\right)\left(\frac{3}{4}\right)\left(\frac{5}{6}\right)\ldots\left(\frac{n-1}{n}\right)\left(\frac{\pi}{2}\right)$$

What's fascinating (or perhaps not?) about this is the integrals hold true is the cosine is now replaced with sine

Now it's easy to just plug these into the formula, but how about we verify some integrals!

Verify:

217) $2\displaystyle\int_0^{\pi/2} \cos^3 x \, dx = \frac{4}{3}$

218) $4\displaystyle\int_0^{\pi/2} \sin^2 x \, dx = \pi$

219) $\displaystyle\int_0^{\pi/2} \cos^7 x \, dx = \frac{16}{35}$

Sine-Cosine with Different Angles

We've come this far without admitting that power reducing integrals are my ultimate kryptonite. They truly scare me and always hate talking about them during LIVES or private tutoring sessions. Eventually, this fear will go away, but what scares me even more are these sons of bitches:

$$A) \int \sin(mx)\sin(nx)\,dx$$

$$B) \int \sin(mx)\cos(nx)\,dx$$

$$C) \int \cos(mx)\cos(nx)\,dx$$

Grrrr....I've seen them on MIT Integration Bee's and when certain teachers assign them to my students. For some reason, they make me quiver.

Nonetheless, here are the formula we should follow:

$$A) = \frac{1}{2}\int \left(\cos[(m-n)x]-\cos[(m+n)x]\right)dx$$

$$B) = \frac{1}{2}\int \left(\sin[(m-n)x]+\sin[(m+n)x]\right)dx$$

$$C) = \frac{1}{2}\int \left(\cos[(m-n)x]+\cos[(m+n)x]\right)dx$$

Save me from these horrendous integrals!

220) $\int \sin(3x) \cos(2x) \, dx$

221) $\int \cos(-3x) \cos(4x) \, dx$

222) $\int \sin(x) \sin(3x) \, dx$

223) $\int \cos(3x) \sin(-4x) \, dx$

Weierstrass Substitution

I know - what a weird name!! But Karl Weierstrass deserves a shoutout for his method we call W-sub

It all begins with integrals involving trig functions. When no u-sub, identity, or power reducing methods help, then w-sub takes over and hopefully gives us an integral recognized using methods discussed in this book.

here's the formula:

$$\int f(\sin x, \cos x)\, dx = \int f\left(\frac{2t}{1+t^2}, \frac{1-t^2}{1+t^2}\right) \frac{2dt}{1+t^2}$$

$$\text{where} \quad t = \tan \frac{x}{2}$$

this is just a fancy way of saying that when we have an integral involving sine and cosine, we can convert every sine and cosine value into the respective values in terms of t and change our dx into the new value on the outside.

This is extremely powerful, and while it can create a head ache of a problem, it's actually solvable.

98

Let's do an example:

$$\text{Take} \quad \int \frac{dx}{1 + \sin x}$$

Here $dx = \frac{2\,dt}{1+t^2}$ and $\sin x = \frac{2t}{1+t^2}$

where $t = \tan\left(\frac{x}{2}\right)$

then $\int \frac{dx}{1+\sin x} = \int \frac{\frac{2\,dt}{1+t^2}}{1 + \frac{2t}{1+t^2}} = \int \frac{2\,dt}{1+t^2+2t}$

$$= \int \frac{2\,dt}{(t+1)^2} = -\frac{2}{t+1} + c$$

since $t = \tan\left(\frac{x}{2}\right) \Rightarrow \boxed{\dfrac{-2}{\tan\left(\frac{x}{2}\right)+1} + c}$

It's easier said than done, but notice that I converted my sine and dx value in terms of t, which turned into a *nicer* integral and re-wrote in terms of x.

Let's see you do some problems of your own. Should you have questions don't hesitate to reach out via Instagram, TikTok, Discord, or email!

224) $\displaystyle\int \frac{dx}{1 + \cos x + \sin x}\, dx$

225) $\displaystyle\int \frac{1}{2 + \cos x}\, dx$

226) $\displaystyle\int \frac{3}{5 - 4\cos x}\, dx$

227) $\displaystyle\int \frac{6}{3 - 2\sin x}\, dx$

228) $\displaystyle\int \frac{1}{1 + \cos 2x}\, dx$

okay, we can start with a u-sub, then w-sub, or maybe use an identify for that cosine?? Up. To. You!

229) $\displaystyle\int \frac{1}{\sec x + 1}\, dx$

230) $\displaystyle\int \frac{1}{\sin(x) + \cos(x)}\, dx$

231) $\displaystyle\int \frac{1}{1 + \csc(x)}\, dx$

232) $\displaystyle\int \frac{1}{\sin^4 x + \cos^4 x}\, dx$

233) $\displaystyle\int \frac{1}{5\sin x + 12\cos x}\, dx$

234) $\displaystyle\int \frac{1}{5\sin x + 2\cos x + 2}\, dx$

235) $\displaystyle\int \frac{1}{4 + 5\cos \frac{x}{2}}\, dx$

The Radicals

I always find the following species of integrals fascinating. The roots can stump many calculus students but knowing the way to break them down and feed them specific u-subs saves us from a lot of heart ache.

These are the forms I am referring to:

1) $\displaystyle\int \frac{1}{(ax+b)\sqrt{px+q}}\,dx$

2) $\displaystyle\int \frac{1}{(ax^2+bx+c)\sqrt{px+q}}\,dx$

3) $\displaystyle\int \frac{1}{(ax+b)\sqrt{px^2+qx+r}}\,dx$

4) $\displaystyle\int \frac{1}{(ax^2+b)\sqrt{px^2+q}}\,dx$

We have reviewed integrals of the form (1) in Page 24, but now we are able to extend it cause we have plenty of practice under our belt. The idea is having some random linear or quadratic function inside the root and a linear or quadratic on the the outside.

Alright, so let's talk about the methods we'll need to use:

1) $\displaystyle\int \frac{1}{(ax+b)\sqrt{px+q}}\, dx$

2) $\displaystyle\int \frac{1}{(ax^2+bx+c)\sqrt{px+q}}\, dx$

For both of these use the following substitution

$$u^2 = px + q$$
$$2u\, du = p\, dx$$

3) $\displaystyle\int \frac{1}{(ax+b)\sqrt{px^2+qx+r}}\, dx$

this one may look funky, but trust me it works!

$$ax + b = \frac{1}{u}$$
$$a\, dx = -\frac{1}{u^2}\, du$$

I want to take this time and shoutout my indian mathematicians since they make great use of this substitution, but instead of u they loooove using t

4) $\displaystyle\int \frac{1}{(ax^2 + b)\sqrt{px^2 + q}}\, dx$

And here's the last substitution to use in these scenarios.

$$x = \frac{1}{u}$$

$$dx = -\frac{1}{u^2}\, du$$

I'd be lying to you if I said these were easy. In my opinion, they are not! Once again, the algebra shows up to call you on your bullshit.

But that does not mean you should be scared and avoid these beasts. Absolutely not!! I encourage you to tackle them, make mistakes, and learn from them. *That* is the only way to grow as a calculus student and mathematician.

Good luck, friend!

236) $\displaystyle\int \frac{1}{(2x + 3)\sqrt{x + 1}}\, dx$

237) $\displaystyle\int \frac{1}{(x - 1)\sqrt{x^2 - 3}}\, dx$

238) $\displaystyle\int \frac{1}{(x+1)^2\sqrt{x^2+2x+2}}\,dx$

239) $\displaystyle\int \frac{1}{(x^2-3x+1)\sqrt{4x+1}}\,dx$

240) $\displaystyle\int \frac{1}{(x-4)\sqrt{x^2-x+3}}\,dx$

241) $\displaystyle\int \frac{1}{(x^2+1)\sqrt{x^2+5}}\,dx$

Sorry, had to throw a wild card in here!

242) $\displaystyle\int \frac{x^2-1}{(x^2+1)\sqrt{x^4+1}}\,dx$

243) $\displaystyle\int \frac{1}{(x+3)\sqrt{x^2-2x+5}}\,dx$

244) $\displaystyle\int \frac{1}{(x^2+3x-9)\sqrt{6x+1}}\,dx$

Area Under the Curve - The Beginning

The purpose of integrals is to accurately find the area underneath the curve. All the methods to solve them is an added bonus, but at the end we always apply the First Fundamental Theorem of Calculus:

$$\int_a^b f(x)\, dx = F(b) - F(a)$$

where F is the anti derivative of f(x) then calculate F(b)-F(a). If you've come this far into the book I am sure that you already knew that, but it doesn't hurt to review a bit.

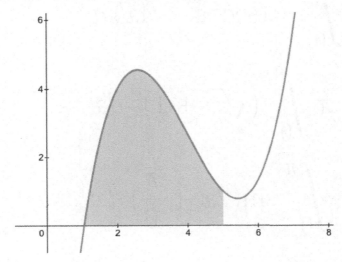

Here are some practice problems to get the wheels turning.

245) $\int_0^4 (x^2 - 3x + 5)\, dx$

246) $\int_{-1}^3 (x + 6\sqrt[3]{x})\, dx$

247) $\int_0^{\pi/2} (\cos x + x)\, dx$

248) $\int_0^{\pi/4} (\sec^2 x - 4x)\, dx$

249) $\pi \int_0^9 (\sqrt{x} + 1)^2\, dx$

250) $\int_0^\pi \sin(2x + \frac{\pi}{6})\, dx$

251) $\int_0^6 \frac{2}{x + 1}\, dx$

We're gonna get to some fun definite integration problems in a bit, but first let's lay out some fundamental concepts worth remembering:

1) If $f(x)$ is an odd function, then

$$\int_{-a}^{a} f(x)\, dx = 0$$

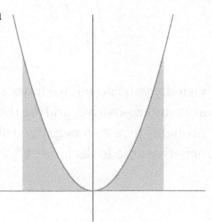

as you can see we have a positive area to the left of the y-axis and a negative on the right. They cancel each other out!

2) If $f(x)$ is an even function, then

$$\int_{-a}^{a} f(x)\, dx = 2 \int_{0}^{a} f(x)\, dx$$

the idea here is splitting the function in half and only focusing on the interval [0,a] and just doubling it.

Sometimes it's easier to deal with 0 than -a

I'd like to continue unraveling this with more practice problems, but trust me when I say it is worth the wait!

For now, let's explore this definite integral:

$$\int_a^b |f(x)|\, dx$$

Very innocent looking and heavily used as the distance of a particle from time a to time b

$$D = \int_a^b |v(t)|\, dt$$

So, just how does the absolute value affect the integration here?

Let's say we are trying to find

$$\int_0^7 (x^2 - 3x - 10)\, dx$$

visually you can tell we have a negative area and a positive, adding those up will probably give us a negative value since the negative area looks largest.

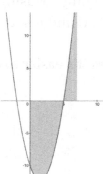

So what happens when we try finding

$$\int_0^7 |x^2 - 3x - 10|\, dx$$

compare this to the previous image and you'll find the negative area is now positive!

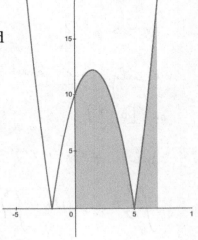

Calculating this is all about finding the x-intercepts of the function, since that is where is changes from negative to positive.

I will show the work on the next page!

110

The trick is to find the x-intercepts and know which portion of the area if negative to make the necessary adjustments!

$$x^2 - 3x - 10 = 0$$

$$(x-5)(x+2) = 0$$

$$x = 5, -2$$

We only care about $x = 5$ since we integrate

on $[0, 7]$

Split up the integral:

$$\int_0^7 (x^2 - 3x - 10) dx = \int_0^5 (x^2 - 3x - 10) dx + \int_5^7 (x^2 - 3x - 10) dx$$

↑

this is the negative

area

$$= -\int_0^5 (x^2 - 3x - 10) dx + \int_5^7 (x^2 - 3x - 10) dx$$

OR

$$= \int_5^0 (x^2 - 3x - 10) dx + \int_5^7 (x^2 - 3x - 10) dx$$

$$= 45\frac{5}{6} + 16\frac{2}{3}$$

$$\boxed{= 62\frac{1}{2}}$$

Let's see you try a few of these problems!

252) $\displaystyle\int_{-2}^{10} |x^2 + 2x - 15|\, dx$

253) $\displaystyle\int_{-2}^{6} |-x^2 + 3x + 1|\, dx$

254) $\displaystyle\int_{0}^{5} |e^{x-2} - 4|\, dx$

255) $\displaystyle\int_{-3}^{5} |\frac{1}{3}x - 2|\, dx$

256) $\displaystyle\int_{-3}^{5} |x^2 - 4x - 8|\, dx$

257) $\displaystyle\int_{0}^{10} |x^2 - 4x - 32|\, dx$

King's Rule

Maybe I am bad at research, but I could not figure out why the following integration technique is called King's Rule, or King's Property. My guess is the one who discovered it (if that is how the story played out) needed the praise.

It's super simple, really, and here it is:

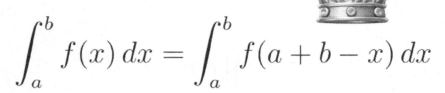

$$\int_a^b f(x)\, dx = \int_a^b f(a+b-x)\, dx$$

Before we continue, here is the proof for those nerds.

$$\text{Let } \int_a^b f(x)\, dx = F(b) - F(a)$$

$$\text{Now } \int_a^b f(a+b-x)\, dx \qquad u = a+b-x$$
$$du = -dx$$

$$\text{if } x = a \rightarrow u = b$$
$$\text{if } x = b \rightarrow u = a$$

$$\text{then } \int_b^a f(u) \cdot -du = -\int_b^a f(u)\, du$$

we can switch the parameters and get

$$\int_a^b f(u)\, du = F(b) - F(a)$$

$$\text{Therefore } \int_a^b f(x)\, dx = \int_a^b f(a+b-x)\, dx$$

This particular property seems very innocent but it can be extremely useful when solving definite integrals involving trig and exponential functions. It makes for some juicy math scribbling!

Let's say we have something that looks like this:

$$\int_0^\pi x\sin^4(x)\,dx$$

there is a possibility we can use integration by parts. I encourage you to try it! But using King's Rule makes this so much easier.

We'll start with labeling the integral as "I" this is important to make some of our steps easier.

$$I = \int_0^\pi x\sin^4(x)\,dx$$

Then apply King's Rule by making another integral "J"

$$J = \int_0^\pi (0+\pi-x)\sin^4(0+\pi-x)\,dx$$

$$J = \int_0^\pi (\pi-x)\sin^4(\pi-x)\,dx$$

$$J = \int_0^\pi (\pi-x)\Big(\sin(\pi)\cos(x)-\cos(\pi)\sin(x)\Big)^4\,dx$$

114

$$J = \int_0^\pi (\pi - x) \sin^4(x)\, dx$$

So far so good, I'd say!

Now look what happens when we add both integrals together - which remember are equal to equal other!

$$I + J = \int_0^\pi x \sin^4(x)\, dx + \int_0^\pi (\pi - x) \sin^4(x)\, dx$$

Since I and J are equal, then I + J = 2I

$$2 \int_0^\pi x \sin^4(x)\, dx = \pi \int_0^\pi \sin^4(x)\, dx$$

or

$$\int_0^\pi x \sin^4(x)\, dx = \frac{\pi}{2} \int_0^\pi \sin^4(x)\, dx$$

now all we have to do it solve the right side, which we can do using power reducing formulas, giving us:

$$\int_0^\pi x \sin^4(x)\, dx = \frac{3\pi^2}{16}$$

In conclusion, when you identity King's Property introduce an "I" for the original integral and make a new "J" integral and combine!

Go ahead and solve the following integrals using the property. Good luck!

258) $\displaystyle\int_0^{\pi/4} \ln(1 + \tan(x))\, dx$

259) $\displaystyle\int_0^{\pi/2} \frac{1}{1 + (\tan(x))^\pi}\, dx$

260) $\displaystyle\int_0^4 \frac{1}{4 + 2^x}\, dx$

261) $\displaystyle\int_0^{\pi/2} \frac{\cos^n(x)}{\sin^n(x) + \cos^n(x)}\, dx$

262) $\displaystyle\int_0^{2\pi} \frac{1}{1 + e^{\sin(x)}}\, dx$

116

263) $\displaystyle\int_0^\pi \frac{x\sin(x)}{1+\cos^2(x)}\,dx$

264) $\displaystyle\int_0^{\pi/2} \ln\left(\frac{1+\sin(x)}{1+\cos(x)}\right)\,dx$

265) $\displaystyle\int_0^\pi \frac{x}{\cos^2(x)+9\sin^2(x)}\,dx$

The S.E-One Method

We're now at the point where some integrals are not as easy to solve as the ones that came before this. For this reason, certain "tricks" are needed. This next one is a special kind of trick that helps solve definite integrals on the interval $[-a, a]$

I call it the S.E - One Method

$$\int_{-a}^{a} \frac{E(x)}{k(x) + 1} = \frac{1}{2} \int_{-a}^{a} E(x)\, dx$$

If you encounter an integral where

S - it's being evaluated on the **same** parameters $-a$ and a
E - there is an **even** function on the numerator
One - the function $k(x)$ on the bottom satisfies $k(x)k(-x) = 1$

then the integral on the right is true and becomes easier to solve.

You will soon find the $k(x)$ is best suited to be an exponential function.

Here is the proof and it involved King's Rule!

Oh what would we do without that King.

$$I = \int_{-a}^{a} \frac{E(x)}{K(x) + 1} \, dx$$

Then $J = \int_{-a}^{a} \frac{E(-a+a-x)}{K(-a+a-x) + 1} \, dx$

$$= \int_{-a}^{a} \frac{E(-x)}{K(-x) + 1} \, dx$$

$E(x) = E(-x)$ since $E(x)$ is even

$$J = \int_{-a}^{a} \frac{E(x)}{K(-x) + 1} \, dx$$

Now multiply $K(x)$

$$J = \int_{-a}^{a} \frac{E(x) \, K(x)}{K(-x)K(x) + 1 \cdot K(x)} \, dx$$

Remember $K(-x) \, K(x) = 1$

$$J = \int_{-a}^{a} \frac{E(x) \, K(x)}{1 + K(x)} \, dx$$

Adding $I + J$

$$2I = \int_{-a}^{a} \frac{E(x)}{K(x)+1} \, dx + \int_{-a}^{a} \frac{E(x) K(x)}{K(x)+1} \, dx$$

$$2I = \int_{-a}^{a} \frac{E(x)(1+K(x))}{(K(x)+1)} \, dx$$

$$2I = \int_{-a}^{a} E(x) \, dx$$

$$I = \frac{1}{2} \int_{-a}^{a} E(x) \, dx$$

All it takes is for you to verify the even function and the special property for $k(x)$ then we are allowed to use this. The proof itself is pretty cool, but implementing it is a bit of a buzzkill as it takes away from the charm of breaking some integrals down.

But that's just me - I want to know what your opinion on this is.

Try some integrals yourself and we'll see what you think.

266) $\displaystyle\int_{-2}^{2} \frac{x^4}{e^x + 1}$

267) $\displaystyle\int_{-\pi}^{\pi} \frac{x\sin(x)}{2^x + 1}$

268) $\displaystyle\int_{-\pi/4}^{\pi/4} \frac{\sqrt{x^2 + 1}}{3^x + 1}$

269) $\pi\displaystyle\int_{-10}^{10} \frac{x^4 + x^2 + 1}{e^x + 1}$

Thoughts? Are these as difficult as the other integrals we have solved, or much easier?

This is where our basic technique exploration ends, unfortunately. There are still numerous methods we can talk about, but since this book is intended for first time, seasoned, and recreational calculus students I don't wish to add too much. The ones mentioned already are *more* than enough for your calculus hearts.

Those looking to get their hands a little dirtier: be on the lookout for more methods and tricks along the way (check table of contents).

More, Please!

Moving forward I will begin to feed you multiple integrals that are worth solving. Not only will they continue to help strengthen your math skills, but - let's be honest - who *doesn't* enjoy solving integrals!

There will be some integrals where I provide a step-by-step solution (in my own scribbles) and briefly introduce methods to approach them.

I've gotta say, having you come along for the journey has been extremely fun. Since my teenage years I dreamt of learning calculus and trying to figure out just what those weird looking S's were. Mathematics has been my passion for years and always dreamt of a day where I can motivate others to love it as much as I do.

I hope that this book (and many others) has done just that.

Mathematics is a universal language that does not change, no matter what corner of the world we stand on. There is no politics here, there is no divide (except maybe on friendly math quarrels), and we should continue to be all accepting.

We should encourage each other to learn, interpret, and teach math in our own unique way. Because there will always be someone out there that needs your explanation for them to move forward.

Thanks for being here, math friend, and Happy Trails

122

270) $\int \dfrac{2x+3}{x^2+2x+10} \, dx$

271) $\int \dfrac{x^2}{\sqrt{25-x^2}} \, dx$

272) $\int \cos^5(x)\sqrt{\sin(x)} \, dx$

273) $\int \dfrac{1}{\sqrt{1+\sin^{-1}x - x^2 - x^2\sin^{-1}x}} \, dx$

I want to give you the opportunity to solve this integral yourself, but check page 124 for my scribbles!

274) $\int \dfrac{1-\tan x}{1+\tan x} \, dx$

275) $\int_0^{\pi/2} \dfrac{\sin(x)}{\sin\left(x+\frac{\pi}{4}\right)} \, dx$

276) $\int e^{3x}\tan^{-1}(e^x) \, dx$

277) $\displaystyle\int \sqrt{(x+1)^2 + 5}\, dx$

278) $\displaystyle\int \frac{1}{x^2 - x - 2}\, dx$

279) $\displaystyle\int \frac{x - 2}{x^2 + 4x - 11}\, dx$

280) $\displaystyle\int \frac{1}{\sqrt{x}(1 + x)}\, dx$

281) $\displaystyle\int \frac{\cos(x)}{\sqrt{4 - \sin^2(x)}}\, dx$

282) $\displaystyle\int \ln(1 + x)\, dx$

283) $\displaystyle\int \frac{1}{\sqrt{x}\,\cos(\sqrt{x})}\, dx$

284) $\displaystyle\int \frac{10}{\sqrt{8x - 4x^2}}\, dx$

124

$$\int \frac{1}{\sqrt{1 + \sin^{-1}x - x^2 - x^2 \sin^{-1}x}} \, dx$$

$$\int \frac{1}{\sqrt{(1-x^2) + \sin^{-1}x\,(1-x^2)}} \, dx$$

$$\int \frac{1}{\sqrt{(1-x^2)(1 + \sin^{-1}x)}} \, dx$$

$$u = 1 + \sin^{-1}x$$

$$du = \frac{1}{\sqrt{1-x^2}} \, dx$$

$$\int u^{-1/2} \, du = 2u^{1/2} + c$$

$$\boxed{= 2\sqrt{1 + \sin^{-1}x} + c}$$

in this particular case I recognized the factoring inside the root and went with a simple u-sub

285) $\displaystyle\int \frac{x+1}{x^2+x-1}\,dx$

286) $\displaystyle\int \frac{1}{x^4+1}\,dx$

okay, I really want to show you this one! Check the next page, but try solving it first!

287) $\displaystyle\int \frac{2}{2\sin^2 x + 4\cos^2 x + 3\sin(2x)}\,dx$

288) $\displaystyle\int \frac{1}{1+\sqrt{3x}}\,dx$

289) $\displaystyle\int \frac{3x^4+2x^3-8x+6}{x^2-1}\,dx$

290) $\displaystyle\int \frac{x+5}{\sqrt{9-(x-3)^2}}\,dx$

291) $\displaystyle\int \tan^5 x\,dx$

126

Problem 286: The fourth polynomial!

$$\int \frac{1}{x^4+1} \, dx = \frac{1}{2} \int \frac{2}{x^4+1} \, dx$$

Do a little re-writing:

$$= \frac{1}{2} \int \frac{(x^2+1) - (x^2-1)}{x^4+1} \, dx$$

and separate:

$$= \frac{1}{2} \int \frac{(x^2+1)}{x^4+1} \, dx - \frac{1}{2} \int \frac{x^2-1}{(x^4+1)} \, dx$$

Divide by x^2!

$$= \frac{1}{2} \int \frac{1 + \frac{1}{x^2}}{x^2 + \frac{1}{x^2}} \, dx - \frac{1}{2} \int \frac{1 - \frac{1}{x^2}}{x^2 + \frac{1}{x^2}} \, dx$$

This is crazy:

$$= \frac{1}{2} \int \frac{1 + \frac{1}{x^2}}{\left(x - \frac{1}{x}\right)^2 + 2} \, dx - \frac{1}{2} \int \frac{1 - \frac{1}{x^2}}{\left(x + \frac{1}{x}\right)^2 - 2} \, dx$$

$$u = x - \frac{1}{x} \qquad\qquad v = x + \frac{1}{x}$$
$$du = 1 + \frac{1}{x^2} \, dx \qquad dv = 1 - \frac{1}{x^2} \, dx$$

$$= \frac{1}{2} \int \frac{1}{u^2+2} \, du - \frac{1}{2} \int \frac{1}{v^2-2} \, dv$$

$$= \frac{1}{2\sqrt{2}} \tan^{-1}\left(\frac{x - \frac{1}{x}}{\sqrt{2}}\right) - \frac{1}{4\sqrt{2}} \ln\left|\frac{\left(x+\frac{1}{x}\right) - \sqrt{2}}{\left(x+\frac{1}{x}\right) + \sqrt{2}}\right| + C$$

292) $\displaystyle\int \sqrt{x^2 + 6x + 1}\, dx$

293) $\displaystyle\int \sqrt{x^2 + 6x + 18}\, dx$

294) $\displaystyle\int \cos^4 x \sin^2 x\, dx$

295) $\displaystyle\int_0^\pi x \sin^5 x\, dx$

296) $\displaystyle\int_0^1 \frac{\ln(1 + x)}{1 + x^2}\, dx$

oooooh I'm *definitely* showing you my scribbles for this one!

297) $\displaystyle\int \frac{\cos(x^3) - 4}{x^2}\, dx$

298) $\displaystyle\int \frac{2x}{x^2 - x - 2}\, dx$

128

Problem 296: Changing to a tangent

$$\int_0^1 \frac{\ln(1+x)}{1+x^2}\,dx$$

Set $x = \tan\theta$ $x=0 \to \theta = 0$

$dx = \sec^2\theta\,d\theta$ $x=1 \to \theta = \pi/4$

$$\int_0^{\pi/4} \frac{\ln(1+\tan\theta)}{1+\tan^2\theta}\cdot \sec^2\theta\,d\theta$$

Remember:

$$1+\tan^2\theta = \sec^2\theta$$

$$\int_0^{\pi/4} \frac{\ln(1+\tan\theta)}{\sec^2\theta}\cdot\sec^2\theta\,d\theta = \int_0^{\pi/4} \ln(1+\tan\theta)\,d\theta$$

King's Rule!

$$I = \int_0^{\pi/4} \ln(1+\tan\theta)\,d\theta$$

$$I = \int_0^{\pi/4} \ln\left(1+\tan\left(\tfrac{\pi}{4}-\theta\right)\right)d\theta = \int_0^{\pi/4}\ln\left(1+\frac{1-\tan\theta}{1+\tan\theta}\right)d\theta$$

$$= \int_0^{\pi/4}\ln\left(\frac{2}{1+\tan\theta}\right)d\theta = \int_0^{\pi/4} \ln(2) - \ln(1+\tan\theta)\,d\theta$$

Now $$I = \int_0^{\pi/4}\ln(2)\,d\theta - \int_0^{\pi/4}\ln(1+\tan\theta)\,d\theta$$

$$2I = \int_0^{\pi/4}\ln(2)\,d\theta = \frac{\pi}{4}\ln(2)$$

$$\boxed{I = \frac{\pi}{8}\ln(2)}$$

299) $\int (5^x - 2^x)^2 \, dx$

300) $\int \dfrac{1}{x^4 - 13x^2 + 36} \, dx$

301) $\int \dfrac{1}{x^9 - x} \, dx$

302) $\int \sqrt{3x^2 + 4x + 5} \, dx$

303) $\int \sqrt{2x^2 + 4x + 5} \, dx$

304) $\int \dfrac{1}{3 + \sin(2x)} \, dx$

305) $\int \dfrac{1}{x} + \dfrac{3}{x^2} - \dfrac{6}{x^3} \, dx$

306) $\displaystyle\int \frac{e^{3x} - 1}{e^x - 1}\, dx$

307) $\displaystyle\int \cos^2\left(\frac{x}{4}\right) dx$

308) $\displaystyle\int \frac{1}{2x + 3}\, dx$

309) $\displaystyle\int \frac{\ln x}{x^2}\, dx$

310) $\displaystyle\int \frac{1}{(x + 3)^2(x - 6)^3}\, dx$

you most certainly can use PFD on this...but check out the neat method introduced on page 132.

311) $\displaystyle\int \frac{1}{(x + 7)^2(x - 8)^3}\, dx$

312) $\displaystyle\int \frac{5x}{\sqrt{4x-x^2}}\, dx$

313) $\displaystyle\int x\sec^{-1}x\, dx$

314) $\displaystyle\int \sin^6 x \cos^5 x\, dx$

315) $\displaystyle\int \frac{1}{x^5 - x^4 + x^3 - x^2 + x - 1}\, dx$

316) $\displaystyle\int \frac{x^3}{(1+x)\sqrt{1+2x-x^2}}\, dx$

317) $\displaystyle\int \frac{4x-2}{x^3-x}\, dx$

318) $\displaystyle\int x\sec x\tan x\, dx$

132

A PFD Trick (sometimes)

Problem 310 has a perfect recipe for a cool trick that I encountered a while back. I don't know what it's called, but if someone out there knows it or would like to come up with something catchy to name it let the world know!

Here is the general form:

$$\int \frac{1}{(x+a)^m (x+b)^n}\, dx$$

This integral is fully prepped for PFD, but when you have this product of linear factors then we can make

$$t = \frac{x+a}{x+b}$$

so that our integral becomes:

$$\frac{1}{(b-a)^{n+m-1}} \int \frac{(1-t)^{n+m-2}}{t^m}\, dt$$

give that a try. Or if you don't want to, then PFD is the way to go!

319) $\int \cos^2(\sin x)\cos x\, dx$

320) $\int_{-4}^{0} 3^{1-x}\, dx$

321) $\int e^x \sqrt{1 + e^{2x}}\, dx$

322) $\int x^2 \tan^{-1}(6x)\, dx$

323) $\int \dfrac{4x - 1}{\sqrt{x^2 + 6x - 11}}\, dx$

324) $\int \dfrac{3x}{x^2 + 1}\, dx$

325) $\int_{0}^{\pi/8} \dfrac{\sin(2x) + 1}{\cos(2x)}\, dx$

326) $\int \dfrac{7xe^{x^2}}{1+e^{2x^2}}\,dx$

327) $\int (4x+7)e^x\,dx$

328) $\int 6\sin^3(x)\sqrt{\cos(2x)}\,dx$

329) $\int_{-\pi/4}^{\pi/4} \dfrac{\cos(x)}{6^x+1}\,dx$

330) $\int_0^1 e^x \sin x\,dx$

331) $\int x^5\sqrt{2-x^3}\,dx$

332) $\int \sqrt{x^2+2x+5}\,dx$

333) $\displaystyle\int \frac{1}{(x+1)^2 + 4}\, dx$

334) $\displaystyle\int \frac{1}{x^2 + 2x + 5}\, dx$

335) $\displaystyle\int \frac{2x^3 - x^2 + 2x + 2}{x^2 + 1}\, dx$

336) $\displaystyle\int_{-5}^{10} |x^2 - 7x - 18|\, dx$

337) $\displaystyle\int \frac{7}{x\sqrt{6x - x^2}}\, dx$

338) $\displaystyle\int \frac{7x^2}{3 - x^3}\, dx$

339) $\displaystyle\int \frac{6x^2 + 3x - 2}{2x}\, dx$

340) $\displaystyle\int \frac{10x}{\sqrt{x^2+1}}\, dx$

341) $\displaystyle\int \frac{1}{\sin x - \sqrt{3}\cos x}\, dx$

342) $\displaystyle\int_0^{25} \frac{1}{\sqrt{4+\sqrt{x}}}\, dx$

343) $\displaystyle\int \sqrt{\tan x}\, dx$

sometimes, integration takes some cleverness. This is one of those. Try solving it first but I will show a solution on Page 138.

344) $\displaystyle\int \frac{x}{(x-1)^6}\, dx$

345) $\displaystyle\int 8^{x^2+2x}(x+1)\, dx$

346) $\displaystyle\int x\cos^2(x^2+1)\, dx$

347) $\displaystyle\int \frac{1}{(x^2 - 2x + 1)(x + 1)^5}\, dx$

348) $\displaystyle\int 4e^x \sin^{-1}(e^x)\, dx$

349) $\displaystyle\int \frac{x}{x^4 + 4x^2 + 8}\, dx$

350) $\displaystyle\int \frac{6}{\sqrt{x}(1 + x)}\, dx$

351) $\displaystyle\int \frac{8}{\sqrt{x}(1 + \sqrt{2x})}\, dx$

352) $\displaystyle\int \frac{5\sec^2 x}{4 + \tan x}\, dx$

353) $\displaystyle\int \sqrt{x^2 + 91}\, dx$

138

Problem 343: root tangent

$$\int \sqrt{\tan x}\ dx \qquad \begin{aligned} u &= \sqrt{\tan x} \\ u^2 &= \tan x \\ 2u\,du &= \sec^2 x\,dx \end{aligned}$$

$$\int u \cdot \frac{2u\,du}{\sec^2 x} = 2\int \frac{u^2}{\sec^2 x}\,du$$

Since $1 + \tan^2 x = \sec^2 x$

$$1 + u^4 = \sec^2 x$$

$$②\int \frac{u^2}{1+u^4}\,du = \int \frac{2u^2}{u^4+1}\,du = \int \frac{u^2+u^2}{u^4+1}\,du$$

$$= \int \frac{(u^2-1) + (u^2+1)}{u^4+1}\,du = \int \frac{u^2-1}{u^4+1}\,du + \int \frac{u^2+1}{u^4+1}\,du$$

Divide by $\frac{1}{u^2} \rightarrow \int \frac{1 - \frac{1}{u^2}}{u^2 + \frac{1}{u^2}}\,du + \int \frac{1 + \frac{1}{u^2}}{u^2 + \frac{1}{u^2}}\,du$

Now the tricky part!
$$\int \frac{1 - \frac{1}{u^2}}{\left(u+\frac{1}{u}\right)^2 - 2}\,du + \int \frac{1 + \frac{1}{u^2}}{\left(u-\frac{1}{u}\right)^2 + 2}\,du$$

Set $\quad w = u + \dfrac{1}{u} \qquad\qquad v = u - \dfrac{1}{u}$

$\quad dw = 1 - \dfrac{1}{u^2}\, du \qquad dv = 1 + \dfrac{1}{u^2}\, du$

now we have:

$$\int \frac{1}{w^2 - 2}\, dw + \int \frac{1}{v^2 + 2}\, dv$$

$$\frac{1}{2\sqrt{2}} \ln \left| \frac{w - \sqrt{2}}{w + \sqrt{2}} \right| + \frac{1}{\sqrt{2}} \arctan\left(\frac{v}{\sqrt{2}}\right) + C$$

$$\frac{1}{2\sqrt{2}} \ln \left| \frac{u + \frac{1}{u} - \sqrt{2}}{u + \frac{1}{u} - \sqrt{2}} \right| + \frac{1}{\sqrt{2}} \arctan\left(\frac{u - \frac{1}{u}}{\sqrt{2}}\right) + C$$

$u = \sqrt{\tan x}$

$$\boxed{\frac{1}{2\sqrt{2}} \ln \left| \frac{\sqrt{\tan x} + \frac{1}{\sqrt{\tan x}} - \sqrt{2}}{\sqrt{\tan x} + \frac{1}{\sqrt{\tan x}} + \sqrt{2}} \right| + \frac{1}{2} \arctan\left(\frac{\sqrt{\tan x} - \frac{1}{\sqrt{\tan x}}}{\sqrt{2}}\right) + C}$$

140

354) $\displaystyle\int \dfrac{1}{e^{4x} + e^{-4x}}\, dx$

355) $\displaystyle\int \dfrac{4x}{\sqrt{-7 - 8x - x^2}}\, dx$

356) $\displaystyle\int x^3 \sin(4x)\, dx$

357) $\displaystyle\int \dfrac{2x + 3}{\sqrt{1 + 2x}}\, dx$

358) $\displaystyle\int \dfrac{\sin(2x)}{\sin^2 x - 2\sin x - 8}\, dx$

359) $\displaystyle\int 3x^2 \tan^{-1}(4x)\, dx$

360) $\displaystyle\int x^2 \ln\left(\dfrac{1 - x}{1 + x}\right) dx$

361) $\displaystyle\int \dfrac{1}{(x^2 + x + 1)\sqrt{x^2 + x + 1}}\, dx$

362) $\int x\sqrt{x^2 - 2x + 2}\,dx$

363) $\int \dfrac{4x - 1}{\sqrt{x^2 + 6x - 11}}\,dx$

364) $\int e^{3x}\sin x\,dx$

This is that ultra-violet voodoo integral, but a perfect opportunity to discuss solving it via Euler's formula. more on page 142.

365) $\int \dfrac{2x^3 + 3x^2 + 4x + 5}{2x^2 + 3x + 4}\,dx$

366) $\int \dfrac{3x + 2}{(x^2 + 4)^{3/2}}\,dx$

367) $\int \dfrac{2x}{x^2 - 4x + 4}\,dx$

368) $\int \dfrac{8x^4}{3 - x^3}\,dx$

369) $\int_{\ln(100)}^{\ln(169)} \dfrac{e^x e^{3x}}{e^{5x} e^{-2x}}\,dx$

Problem 364: Euler's Formula

The following method is brought up from students looking to flex on their knowledge of euler's formula. And they deserve to - this formula is incredibly neat, powerful, and used immensely in higher level mathematics.

First, shoutout to Euler. The more I learn about his work the more I'm convinced he just might be the greatest mathematician of all time. The real GOAT.

bow down, bitches!

This is his formula:

$$e^{ix} = \cos(x) + i\sin(x)$$

It's derived using a MacLaurin Series for e and rearranged a series of cosines, sines, and imaginary numbers.

Now we're able to call for the Real Value of this formula to give us cosine:

$$Re(e^{ix}) = Re(\cos(x) + i\sin(x)) = \cos(x)$$

and if we wanted sine (not the imaginary part of sine!) We'll first have to do:

$$-ie^{ix} = -i\cos(x) - i^2\sin(x)$$
$$-ie^{ix} = \sin(x) - i\cos(x)$$
$$Re(-ie^{ix}) = Re(\sin(x) - i\cos(x)) = \sin(x)$$

this will help us solve: $\displaystyle\int e^{3x} \sin x \, dx$

$$= \int e^{3x} \, Re(-i \, e^{ix}) \, dx$$ ← here we call for sine

$$= \int Re(-i \, e^{ix+3x}) \, dx$$ and use properties of exponents

$$= \int Re(-i \, e^{x(i+3)}) \, dx$$ 🌙

$$= Re\left(\frac{-i}{i+3} \, e^{x(i+3)}\right) + C$$ we integrate e

$$\frac{-i}{i+3} = -\frac{1}{10} - \frac{3}{10} i$$ and find the value of this complex expression

$$= Re\left(\left(-\frac{1}{10} - \frac{3}{10}i\right) e^{ix} \cdot e^{3x}\right) + C$$ multiply and simplify!

$$= Re\left(\left(-\frac{1}{10} - \frac{3}{10}i\right)(\cos x + i \sin x) e^{3x}\right) + C$$ ↘

$$= Re\left(\left(-\frac{1}{10}\cos x - \frac{1}{10}i\sin x - \frac{3}{10}i\cos x + \frac{3}{10}\sin x\right)e^{3x}\right) + C$$

$$= \left(-\frac{1}{10}\cos x + \frac{3}{10}\sin x\right)e^{3x} + C$$

144

370) $\int \dfrac{3x-1}{4x^2-4x+17}\,dx$

371) $\int \dfrac{\sin x - 2\cos x}{1+4\sin x \cos x}\,dx$

372) $\int e^{7x}\cos x\,dx$

373) $\int e^{x}\cos(7x)\,dx$

try and see if you can use Euler's formula on the previous two!

374) $\int \dfrac{(x^4+1)}{x^6+1}\,dx$

375) $\int \dfrac{1}{6x^4-11x^3-51x^2-6x+8}\,dx$

376) $\int \dfrac{\tan^{-1}x}{(x-1)^3}\,dx$

377) $\int (5^x - 2^x)^2\,dx$

378) $\displaystyle\int \frac{1}{\sin^2(6x)}\, dx$

379) $\displaystyle\int x^2 \ln(1+x)\, dx$

380) $\displaystyle\int \sqrt{(x+3)(x+4)(x+5)(x+6)+1}\, dx$

$$x \geq -3$$

I will *have* to address the controversy on this problem! Check Page 146.

381) $\displaystyle\int \frac{1}{\sqrt{3x+4}+\sqrt{2x+1}}\, dx$

382) $\displaystyle\int \frac{4}{6\sin^2 x + 5}\, dx$

Hint: divide everything by cos^2

383) $\displaystyle\int \frac{1}{x^2 + 6x + 19}\, dx$

384) $\displaystyle\int \frac{1}{\sqrt{x^2 + 6x + 19}}\, dx$

146

Problem 380: the quad-root trick

I found this in an another book of integrals and thought this trick was just...fun. I made a few videos about it and the comments either praised me, or dragged me.

One thing a follower mentioned was that he "already knew the product of four consecutive integers plus 1 was a perfect square." I set out to prove that (which was fun!) and realized just how easy this quad-trick can be! Here it goes:

$$\int \sqrt{(x+3)(x+4)(x+5)(x+6) + 1} \; dx$$

$$\int \sqrt{(x+3)(x+6)(x+4)(x+5) + 1} \; dx$$

$$\int \sqrt{(x^2+9x+18)(x^2+9x+20) + 1} \; dx$$

Make $\alpha = x^2 + 9x$

it starts off by rearranging and multiplying the proper linear factors. Then we recognize that they both share the alpha value!

$$\int \sqrt{(\alpha + 18)(\alpha + 20) + 1} \; dx$$

$$\int \sqrt{\alpha^2 + 38\alpha + 360 + 1} \; dx$$

$$\int \sqrt{\alpha^2 + 38\alpha + 361} \; dx$$

$$\int \sqrt{(\alpha + 19)^2} \; dx$$

$$\int \alpha + 19 \; dx$$

Here is the controversy! Since the root is removed, should my answer include a plus/minus or not?? You tell me.

$$\int x^2 + 9x + 19 \; dx$$

$$= \frac{x^3}{3} + \frac{9x^2}{2} + 19x + C$$

148

385) $\int \sqrt{x^2 + 3x - 1}\, dx$

386) $\int 2\sin(2x)(1 - \cos(2x))\, dx$

387) $\int 6\sin(x)e^{\cos(x)+6}\, dx$

388) $\int \dfrac{3}{6x + 7}\, dx$

389) $\int \dfrac{3x^3}{1 + 6x^2}\, dx$

390) $\int \dfrac{x(x + 2)}{x^3 + 3x^2 - 4}\, dx$

391) $\int \dfrac{1 + x}{\sqrt{81 - x^2}}\, dx$

392) $\int 6\sin^3(3x)\, dx$

393) $\displaystyle\int_{\pi/2}^{\pi} e^{x}\left(\dfrac{1-\sin x}{1-\cos x}\right) dx$

394) $\displaystyle\int_{0}^{\pi/2} \ln(\sec x)\, dx$

this was brought to my attention by someone on instagram (you know who you are!) and allow me to give me solution on the next page!

395) $\displaystyle\int_{0}^{1} \dfrac{1}{\sqrt{1+x}-\sqrt{x}}\, dx$

396) $\displaystyle\int x^{4}\sin(2x)\, dx$

397) $\displaystyle\int \sqrt{x\sqrt{x\sqrt{x\sqrt{x\ldots}}}}\, dx$

398) $\displaystyle\int_{0}^{\pi/4} \dfrac{\sin x + \cos x}{9 + 16\sin(2x)}\, dx$

399) $\displaystyle\int \dfrac{\tan^{-1}(3x)}{x^{2}}\, dx$

150

Problem 394: a wild ride

I first attempted this using King's Rule but nothing to seemed to work (boo!) So, we take the scenic route.

$$\int_0^{\pi/2} \ln\left(\sec(x)\right) dx$$

Since $\sec(x) = \dfrac{1}{\cos(x)}$ then

$$\int_0^{\pi/2} \ln\left(\sec(x)\right) dx = \int_0^{\pi/2} \ln\left(\frac{1}{\cos(x)}\right) dx$$

$$= \int_0^{\pi/2} \ln(1) - \ln\left(\cos(x)\right) dx$$

$$= -\int_0^{\pi/2} \ln\left(\cos(x)\right) dx$$

Trust me, this is for our own good!

Now we have $I = -\int_0^{\pi/2} \ln(\cos(x)) dx$

and we can time travel and get

$$I = -2\int_0^{\pi/4} \ln(\cos(2x)) dx$$

Notice the changes ... do a u-sub to check my work!

$$I = -2\int_0^{\pi/4} \ln(\cos(2x)) dx$$

Since $\cos(2x) = \cos^2(x) - \sin^2(x)$

$$I = -2\int_0^{\pi/4} \ln(\cos^2(x) - \sin^2(x)) dx$$

and bring in difference of squares

$$I = -2\int_0^{\pi/4} \ln((\cos(x) - \sin(x))(\cos(x) + \sin(x))) dx$$

and use properties of logarithms

$$I = -2\int_0^{\pi/4} \ln(\cos(x) - \sin(x)) dx - 2\int_0^{\pi/4} \ln(\cos(x) + \sin(x)) dx$$

152

$$I = -2 \int_0^{\pi/4} \ln\left(\cos(x) - \sin(x)\right) dx - 2 \int_0^{\pi/4} \ln\left(\cos(x) + \sin(x)\right) dx$$

Okay, hear me out:

$$\cos\left(x + \tfrac{\pi}{4}\right) = \cos(x)\cos\left(\tfrac{\pi}{4}\right) - \sin(x)\sin\left(\tfrac{\pi}{4}\right)$$

$$\cos\left(x + \tfrac{\pi}{4}\right) = \tfrac{\sqrt{2}}{2}\cos(x) - \tfrac{\sqrt{2}}{2}\sin(x)$$

$$\sqrt{2}\cos\left(x + \tfrac{\pi}{4}\right) = \cos(x) - \sin(x)$$

$$\cos\left(x - \tfrac{\pi}{4}\right) = \cos(x)\cos\left(\tfrac{\pi}{4}\right) + \sin(x)\sin\left(\tfrac{\pi}{4}\right)$$

$$\cos\left(x - \tfrac{\pi}{4}\right) = \tfrac{\sqrt{2}}{2}\cos(x) + \tfrac{\sqrt{2}}{2}\sin(x)$$

$$\sqrt{2}\cos\left(x - \tfrac{\pi}{4}\right) = \cos(x) + \sin(x)$$

So, $$I = -2\int_0^{\pi/4} \ln\left(\sqrt{2}\cos\left(x + \tfrac{\pi}{4}\right)\right) dx - 2\int_0^{\pi/4} \ln\left(\sqrt{2}\cos\left(x - \tfrac{\pi}{4}\right)\right) dx$$

Using more log properties!

$$I = -2\int_0^{\pi/4} \ln(\sqrt{2})\, dx - 2\int_0^{\pi/4} \ln\left(\cos\left(x + \tfrac{\pi}{4}\right)\right) dx$$

$$\qquad -2\int_0^{\pi/4} \ln(\sqrt{2})\, dx - 2\int_0^{\pi/4} \ln\left(\cos\left(x - \tfrac{\pi}{4}\right)\right) dx$$

Since $-2\int_0^{\pi/4} \ln(\sqrt{2})\, dx = -2\left[x\ln(\sqrt{2})\right]_0^{\pi/4}$

$$= -2\left[\frac{\pi}{4}\ln(\sqrt{2})\right]$$

$$= -\frac{\pi}{2}\ln(\sqrt{2})$$

$$= -\frac{\pi}{4}\ln(2)$$

$I = -\frac{\pi}{4}\ln(2) - 2\int_0^{\pi/4} \ln\left(\cos\left(x+\frac{\pi}{4}\right)\right) dx$

$$-\frac{\pi}{4}\ln(2) - 2\int_0^{\pi/4} \ln\left(\cos\left(x-\frac{\pi}{4}\right)\right) dx$$

$I = -\frac{\pi}{2}\ln(2) - 2\int_0^{\pi/4} \ln\left(\cos\left(x+\frac{\pi}{4}\right)\right) dx - 2\int_0^{\pi/4} \ln\left(\cos\left(x-\frac{\pi}{4}\right)\right) dx$

$I = -\frac{\pi}{2}\ln(2) - 2\int_0^{\pi/4} \ln\left(\cos\left(x+\frac{\pi}{4}\right)\cos\left(x-\frac{\pi}{4}\right)\right) dx$

Here $\cos\left(x+\frac{\pi}{4}\right)\cos\left(x-\frac{\pi}{4}\right) = \dfrac{\cos(2x)}{2}$

$I = -\frac{\pi}{2}\ln(2) - 2\int_0^{\pi/4} \ln\left(\frac{\cos(2x)}{2}\right) dx$

$I = -\frac{\pi}{2}\ln(2) - 2\int_0^{\pi/2} \ln\left(\cos(x)\right) dx$

↖ Look! It's

$-I$

$$I = -\frac{\pi}{2} \ln(2) + 2I$$

$$-I = -\frac{\pi}{2} \ln(2)$$

$$\boxed{I = \frac{\pi}{2} \ln(2)}$$

...what just happened?

400 $\displaystyle\int (1-x)(1-2x)(1-3x)\,dx$

401) $\displaystyle\int \frac{x^2}{(1-x)^{100}}\,dx$

402) $\displaystyle\int \frac{\sin x \cos x}{\sqrt{3 - \sin^4 x}}\,dx$

403) $\displaystyle\int \frac{x^4 - 5x^3 + 6x^2 - 18}{x^3 - 3x^2}\,dx$

404) $\displaystyle\int_0^1 \sin^2(\ln x)\, dx$

405) $\displaystyle\int \frac{1}{x^5 + 1}\, dx$

okay, I'm ready to talk about this! Check out the next page.

406) $\displaystyle\int \frac{x^2 + 4}{3x^3 + 4x^2 - 4x}\, dx$

407) $\displaystyle\int \frac{6x - 1}{4x^2 - 8x + 1}\, dx$

408) $\displaystyle\int \frac{1}{x^2 + 2x + 5}\, dx$

409) $\displaystyle\int_{-5}^5 x\sqrt{x^2 + \cos x}\, dx$

Problem 405: Demoivre's Theorem

I have seen some solutions to this particular integral, but Demoivre's Theorem calls to me whenever I see $\displaystyle\int \frac{1}{x^5 + 1}\, dx$

First, let's give it up for Abraham de Moivre. Thanks to his contributions we are able to use a formula that helps us find the complete complex solutions to an equation of higher degree.

Merci beaucoup!!

His formula tells us if we have a complex number in trigonometric form

$$z = r(\cos\theta + i\sin\theta)$$

then we can find the powers of these complex numbers connecting them to the angle formed on the complex plane:

$$z^n = r^n(\cos n\theta + i\sin n\theta)$$

or better yet, take their roots

$$z^{1/n} = r^{1/n}\left(\cos\frac{\theta}{n} + i\sin\frac{\theta}{n}\right)$$

So how does this all help us solve for $\int \dfrac{1}{x^5 + 1}\, dx$??

First, let's break the denominator down. Let me remind you that

$$e^{ix} = \cos(x) + i\sin(x)$$

thank you - thank you Mr. Euler! Now we can find all the factors of the denominator by finding the value of -1 using Euler's formula and finding the fifth roots by Mr. DeMoivre's!

$$x^5 + 1 = 0$$

$$x^5 = -1$$

Since $x^5 = e^{i(\pi + 2\pi k)}$ by Eulee's

then $x = \sqrt[5]{e^{i(\pi + 2\pi k)}}$ by Demoivre's

$$x = e^{i\left(\frac{\pi + 2\pi k}{5}\right)}$$

$$x = e^{i\left(\frac{\pi}{5} + \frac{2\pi k}{5}\right)}$$

Now let's find all the propers values for k.

$$K = 0 \quad \rightarrow \quad X = e^{i\pi/5}$$

$$K = 1 \quad \rightarrow \quad X = e^{i\,3\pi/5}$$

$$K = 2 \quad \rightarrow \quad X = e^{i\pi}$$

$$K = 3 \quad \rightarrow \quad X = e^{i\,7\pi/5}$$

$$K = 4 \quad \rightarrow \quad X = e^{i\,9\pi/5}$$

$$\int \frac{1}{x^5+1}\,dx = \int \frac{1}{\left(x-e^{i\frac{\pi}{5}}\right)\left(x-e^{i\frac{3\pi}{5}}\right)\left(x-e^{i\pi}\right)\left(x-e^{i\frac{7\pi}{5}}\right)\left(x-e^{i\frac{9\pi}{5}}\right)}\,dx$$

Given these values we can now re-write our integral using these factors and get to using PFD.

$$\frac{A}{\left(x-e^{i\frac{\pi}{5}}\right)} + \frac{B}{\left(x-e^{i\frac{3\pi}{5}}\right)} + \frac{C}{\left(x-e^{i\pi}\right)} + \frac{D}{\left(x-e^{i\frac{7\pi}{5}}\right)} + \frac{E}{\left(x-e^{i\frac{9\pi}{5}}\right)}$$

$$A\left(x-e^{i\frac{3\pi}{5}}\right)\left(x-e^{i\pi}\right)\left(x-e^{i\frac{7\pi}{5}}\right)\left(x-e^{i\frac{9\pi}{5}}\right)$$

$$+ B\left(x-e^{i\frac{\pi}{5}}\right)\left(x-e^{i\pi}\right)\left(x-e^{i\frac{7\pi}{5}}\right)\left(x-e^{i\frac{9\pi}{5}}\right)$$

$$+ C\left(x-e^{i\frac{\pi}{5}}\right)\left(x-e^{i\frac{3\pi}{5}}\right)\left(x-e^{i\frac{7\pi}{5}}\right)\left(x-e^{i\frac{9\pi}{5}}\right)$$

$$+ D\left(x-e^{i\frac{\pi}{5}}\right)\left(x-e^{i\frac{3\pi}{5}}\right)\left(x-e^{i\pi}\right)\left(x-e^{i\frac{9\pi}{5}}\right)$$

$$+ E\left(x-e^{i\frac{\pi}{5}}\right)\left(x-e^{i\frac{3\pi}{5}}\right)\left(x-e^{i\pi}\right)\left(x-e^{i\frac{7\pi}{5}}\right)$$

$$= 1$$

Now we find the values for A, B, C, D, and E

$$X = e^{i\frac{\pi}{5}} \rightarrow A = \cfrac{1}{\left(e^{i\frac{\pi}{5}}-e^{i\frac{3\pi}{5}}\right)\left(e^{i\frac{\pi}{5}}-e^{i\pi}\right)\left(e^{i\frac{\pi}{5}}-e^{i\frac{7\pi}{5}}\right)\left(e^{i\frac{\pi}{5}}-e^{i\frac{9\pi}{5}}\right)}$$

$$X = e^{i\frac{3\pi}{5}} \rightarrow B = \cfrac{1}{\left(e^{i\frac{3\pi}{5}}-e^{i\frac{\pi}{5}}\right)\left(e^{i\frac{3\pi}{5}}-e^{i\pi}\right)\left(e^{i\frac{3\pi}{5}}-e^{i\frac{7\pi}{5}}\right)\left(e^{i\frac{3\pi}{5}}-e^{i\frac{9\pi}{5}}\right)}$$

$$X = e^{i\pi} \rightarrow C = \cfrac{1}{\left(e^{i\pi}-e^{i\frac{\pi}{5}}\right)\left(e^{i\pi}-e^{i\frac{3\pi}{5}}\right)\left(e^{i\pi}-e^{i\frac{7\pi}{5}}\right)\left(e^{i\pi}-e^{i\frac{9\pi}{5}}\right)}$$

$$X = e^{i\frac{7\pi}{5}} \rightarrow D = \cfrac{1}{\left(e^{i\frac{7\pi}{5}}-e^{i\frac{\pi}{5}}\right)\left(e^{i\frac{7\pi}{5}}-e^{i\frac{3\pi}{5}}\right)\left(e^{i\frac{7\pi}{5}}-e^{i\pi}\right)\left(e^{i\frac{7\pi}{5}}-e^{i\frac{9\pi}{5}}\right)}$$

$$X = e^{i\frac{9\pi}{5}} \rightarrow E = \cfrac{1}{\left(e^{i\frac{9\pi}{5}}-e^{i\frac{\pi}{5}}\right)\left(e^{i\frac{9\pi}{5}}-e^{i\frac{3\pi}{5}}\right)\left(e^{i\frac{9\pi}{5}}-e^{i\pi}\right)\left(e^{i\frac{9\pi}{5}}-e^{i\frac{7\pi}{5}}\right)}$$

and we finish it off!

$$\int \frac{A}{x-e^{i\frac{\pi}{5}}}\,dx + \int \frac{B}{x-e^{i\frac{3\pi}{5}}}\,dx + \int \frac{C}{x-e^{i\pi}}\,dx + \int \frac{D}{x-e^{i\frac{7\pi}{5}}}\,dx + \int \frac{E}{x-e^{i\frac{9\pi}{5}}}\,dx$$

$$= A \ln\left|x-e^{i\frac{\pi}{5}}\right| + B \ln\left|x-e^{i\frac{3\pi}{5}}\right| + C \ln\left|x-e^{i\pi}\right|$$
$$+ D \ln\left|x-e^{i\frac{7\pi}{5}}\right| + E \ln\left|x-e^{i\frac{9\pi}{5}}\right| + C$$

410) $\int \cot x (1 - \cos(2x))\, dx$

411) $\int \dfrac{5x + 1 - x^3}{e^{2x}}\, dx$

412) $\int 4x e^{x} \sin x\, dx$

413) $\int \sqrt{3x - x^2}\, dx$

414) $\int_0^{\pi/2} \dfrac{\cos^2 x}{\cos^2 x + 4\sin^2 x}\, dx$

415) $\int \dfrac{2x^3 - x^2 + 4x + 34}{x^4 - 6x^2 - 8x + 24}\, dx$

416) $\int \dfrac{2x^3 - 2x^2 + 8x + 30}{x^4 - 6x^2 - 8x + 24}\, dx$

417) $\displaystyle\int_0^1 \frac{x^2-1}{\ln x}\,dx$

I am very excited to show and talk about the following technique. Trust me, you won't want to miss Page 162.

418) $\displaystyle\int \frac{1}{\sqrt{x^2+1}-\sqrt{x^2-1}}\,dx$

419) $\displaystyle\int \frac{1}{x^4-1}\,dx$

420) $\displaystyle\int \frac{x^{2n-1}}{x^n+1}\,dx$

421) $\displaystyle\int \frac{4x}{\sqrt{5x+1}}\,dx$

422) $\displaystyle\int (2+5\cos x)^3 \sin x\,dx$

162

Remember the scene from *Forrest Gump* where Forrest is running on a highway overlooking Monument Valley? He's been running for thousands of miles, gathering other runners to accompany him on the adventure, and he finally comes to a halt.

He turns to face the group of runners, ponders for a second or two and says

"I'm kinda tired...I'd like to go home now."

That is the sentiment. Not only has it been a journey writing and scribbling, but I can only imagine how you as the reader feels. I think it's time we start wrapping up.

So, the next couple pages will be filled with 3 specific topics: the Feynman technique, Taylor/MacLaurin Series, and close it off with introducing the dilogarithm function.

If you're looking to explore more, come along with me! If you're just taking calculus and feel fulfilled with these integrals, trust me when I say this is a good stopping point.

Problem 417: The Feynman Technique

I remember reading *Surely you're joking, Mr Feynman?* when I was in high school and was just fascinated with Feynman's way of thinking, his quirkiness, and his all around knowledge of the world. I cannot say better things about this man.

It wasn't until recent years that I began understanding his technique to solving integrals, which he talked about in his book!

The idea is to set the integral as a function of another parameter and "differentiate under the integral sign." This comes down to taking the partial derivative with respect to the new parameter we introduced.

This will give us an "easier" integral to work with and then bring it back by integrating with respect to the new parameter, thus giving us the solution to our integral.

Let's talk about this technique using:

$$\int_0^1 \frac{x^2 - 1}{\ln x}\, dx$$

164

We'll start by generalizing the integral for some new parameter:

$$f(\alpha) = \int_0^1 \frac{x^\alpha - 1}{\ln x} \, dx$$

Since we can make alpha whatever we want, at the end we are concerned with the $f(2)$

One thing I want to point out right now. If we make alpha 0 we get:

$$f(0) = \int_0^1 \frac{x^0 - 1}{\ln x} \, dx = \int_0^1 \frac{1 - 1}{\ln x} \, dx = 0$$

we will revisit this later!

Back to our integral. This is the feynman technique: take the derivative under the integral sign, or in other words, take the *partial* derivative with respect to alpha.

For those that are not familiar with the partial derivative (which is why I didn't talk about this till the end) you essentially treat all the other variables like a constant and only focus on the variable we are differentiating with respect to.

$$\frac{\partial}{\partial \alpha} f(\alpha) = \frac{\partial}{\partial \alpha} \int_0^1 \frac{x^\alpha - 1}{\ln x} \, dx$$

$$\frac{\partial}{\partial \alpha} f(\alpha) = \int_0^1 \frac{\partial}{\partial \alpha} \frac{x^\alpha - 1}{\ln x} \, dx$$

$$\frac{\partial}{\partial \alpha} f(\alpha) = \int_0^1 \frac{x^\alpha \ln x}{\ln x} \, dx$$

$$f'(\alpha) = \int_0^1 x^\alpha \, dx$$

look how much easier this becomes. Jeez, this seriously feels like magic. We'll now focus on the right hand side we can integrate with respect to x. In this case alpha is some random integer.

$$f'(\alpha) = \left[\frac{x^{\alpha+1}}{\alpha + 1} \right]_0^1$$

$$f'(\alpha) = \frac{(1)^{\alpha+1}}{\alpha + 1} - \frac{(0)^{\alpha+1}}{\alpha + 1}$$

$$f'(\alpha) = \frac{1}{\alpha + 1}$$

Well will you look at this! Using first fundamental theorem of calculus gives us a pretty single answer. But this is f', let's find the original function.

166

$$\int f'(\alpha)\, d\alpha = \int \frac{1}{\alpha + 1}\, d\alpha$$

$$f(\alpha) = \ln |\alpha + 1| + C$$

...but what is the value of C? Remember

$$f(0) = \int_0^1 \frac{x^0 - 1}{\ln x}\, dx = \int_0^1 \frac{1 - 1}{\ln x}\, dx = 0$$

that must mean that

$$f(0) = \ln |0 + 1| + C$$

$$0 = C$$

Which leave us with:

$$f(\alpha) = \int_0^1 \frac{x^\alpha - 1}{\ln x} = \ln |\alpha + 1|$$

$$f(2) = \int_0^1 \frac{x^2 - 1}{\ln x} = \ln |2 + 1| = \ln(3)$$

Mind blown! This technique is so fucken cool, but takes time to perfect. First, one must be able to identity when to use it, where to put the alpha parameter, and then taking care of the strangling constant that comes with it.

Give it a try with the following:

423) $\displaystyle\int_0^\infty \frac{\tan^{-1}(\pi x) - \tan^{-1}(2x)}{x}\, dx$

using

$$f(\alpha, \beta) = \int_0^\infty \frac{\tan^{-1}(\alpha x) - \tan^{-1}(\beta x)}{x}\, dx$$

424) $\displaystyle\int_0^\infty \frac{\ln x}{1 + x^2}\, dx$

using

$$f(\alpha) = \int_0^\infty \frac{\ln \alpha x}{1 + x^2}\, dx$$

425) $\displaystyle\int_0^\infty \frac{\sin x}{x}\, dx$

using

$$f(\alpha) = \int_0^\infty \frac{e^{-\alpha x} \sin x}{x}\, dx$$

426) $\displaystyle\int_0^\pi \ln(5 - 4\cos x)\, dx$

using

$$f(\alpha) = \int_0^\pi \ln(1 - 2\alpha \cos x + \alpha^2)\, dx$$

Taylor (no, not *that* Taylor) Expansion

I will not dare badmouth t-swift, include any pics, or defame her in any way...not only will she suck every penny out of me, but her swifties will ultimately drag me to the ground harder than any integral encountered.

But if somehow, in an another dimension, T-Swift was a famous mathematician that just won an award, I'd storm the stage as she stood there, grab the mic out of her hand and claim, "Taylor, I'm really happy for you, but Brook Taylor had one of the best mathematical findings of all time!"

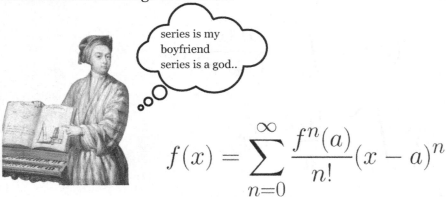

series is my boyfriend series is a god..

$$f(x) = \sum_{n=0}^{\infty} \frac{f^n(a)}{n!}(x-a)^n$$

I have two more integration techniques that will require knowledge of Taylor (more specifically MacLaurin) series. The more you study these the more you realize that, shit, *everything* is a series.

That's right, every single function we integrated previously can *technically* be written as a series.

The concept is quite simple. Given a continuous and differentiable function $f(x)$, this particular function can be written as a series of its derivatives evaluated at a value a, which we say is centered at that point.

$$f(x) = \sum_{n=0}^{\infty} \frac{f^n(a)}{n!}(x-a)^n$$

When the function is centered at a = 0 our series becomes

$$f(x) = \sum_{n=0}^{\infty} \frac{f^n(0)}{n!}x^n$$

We call this a MacLaurin series, which is what we're going to work with. Taylor generalized it, MacLaurin gave it a deeper meaning.

I encourage you to look up how to find Maclaurin series of functions. Part of learning mathematical concepts is exploring them via different avenues. Ask your teacher, find a video online, or find a textbook explaining it.

Believe it or not, some series are better to work with than the actual functions, just like most functions are better to work with than its series counterpart. It's our job to find out when it's better to work with one vs the other.

Before we continue let me introduce some important series that not only come up often in our problems, but their series are easy to find!

$$\frac{1}{1-x} = 1 + x + x^2 + x^3 + \ldots = \sum_{n=0}^{\infty} x^n$$

$$e^x = 1 + x + \frac{x^2}{2!} + \frac{x^3}{3!} + \ldots = \sum_{n=0}^{\infty} \frac{x^n}{n!}$$

$$\sin x = x - \frac{x^3}{3!} + \frac{x^5}{5!} - \frac{x^7}{7!} + \ldots = \sum_{n=0}^{\infty} \frac{(-1)^n x^{2n+1}}{(2n+1)!}$$

$$\cos x = 1 - \frac{x^2}{2!} + \frac{x^4}{4!} - \frac{x^6}{6!} + \ldots = \sum_{n=0}^{\infty} \frac{(-1)^n x^{2n}}{(2n)!}$$

these are four of the basic maclaurin series and there are two more I want to show you. However, we will use our known series to find them. I will show you what I mean in a second. But first I want to talk about

$$e^{ix} = \cos(x) + i\sin(x)$$

Euler's formula has three series we now know. Check out the next page to see how we can derive this formula using their Maclaurin series.

We begin this journey with exp(x) and now introduced the complex
number i into the mix...

$$e^x = 1 + x + \frac{x^2}{2!} + \frac{x^3}{3!} + \frac{x^4}{4!} + \frac{x^5}{5!} + \frac{x^6}{6!} + \frac{x^7}{7!} + \cdots$$

$$e^{ix} = 1 + (ix) + \frac{(ix)^2}{2!} + \frac{(ix)^3}{3!} + \frac{(ix)^4}{4!} + \frac{(ix)^5}{5!} + \frac{(ix)^6}{6!} + \frac{(ix)^7}{7!} + \cdots$$

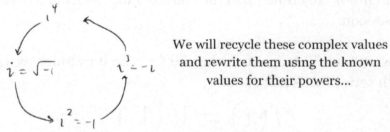

We will recycle these complex values
and rewrite them using the known
values for their powers...

$$e^{ix} = 1 + ix - \frac{x^2}{2!} - \frac{ix^3}{3!} + \frac{x^4}{4!} + \frac{ix^5}{5!} - \frac{x^6}{6!} - \frac{ix^7}{7!} + \cdots$$

$$e^{ix} = \left(1 - \frac{x^2}{2!} + \frac{x^4}{4!} - \frac{x^6}{6!} + \cdots \right) + i\left(x - \frac{x^3}{3!} + \frac{x^5}{5!} - \frac{x^7}{7!} + \cdots \right)$$

and then we group the reals and imaginaries together (which
team are you??)

Recognize these separate series???

$$e^{ix} = \sum_{n=0}^{\infty} \frac{(-1)^n x^{2n}}{(2n)!} + i \sum_{n=0}^{\infty} \frac{(-1)^n x^{2n+1}}{(2n+1)!}$$

$$e^{ix} = \cos(x) + i \sin(x)$$

172

I still remember my calculus teacher, Mr. Bolen, showing us this. At the time I was into rubik's cubes and was focused on figuring out an algorithm. Admittedly, I was *not* paying attention and Mr. Bolen knew that. So he yells,

"Miguel, you don't want to miss this!"

Well, good thing I looked up and saw euler's formula being derived by its series cause ever since then I could not stop writing this "proof" any time I had the chance. Euler's formula became an obsession.

It wasn't until I taught calculus did I realize how much we can do with series! Let's talk about

$$f(x) = \ln(1 + x)$$

we can do all the work of taking multiple derivatives to figure out its series, but trust me it's such a struggle! The second and third derivative is already a pain in the ass, so why go through all the trouble!

Instead, we will work with the first series I introduced on page 169. From there we change the value of x and integrate, giving us the series for this function.

Check it out

$$\frac{1}{1-x} = 1+x+x^2+x^3+\cdots = \sum_{n=0}^{\infty} x^n$$

$$\frac{1}{1-(-x)} = 1+(-x)+(-x)^2+(-x)^3+\cdots = \sum_{n=0}^{\infty} (-x)^n$$

$$\frac{1}{1+x} = 1-x+x^2-x^3+\cdots = \sum_{n=0}^{\infty} (-1)^n x^n$$

$$\int \frac{1}{1+x}\,dx = \int (1-x+x^2-x^3+\cdots)\,dx = \sum_{n=0}^{\infty} \int (-1)^n x^n\,dx$$

$$\ln(1+x) = x-\frac{x^2}{2}+\frac{x^3}{3}-\frac{x^4}{4}+\cdots = \sum_{n=0}^{\infty} \frac{(-1)^n x^{n+1}}{n+1}$$

whoa - cheat code unlocked!! This is already making our life ALOT easier.

Not only are we able to find series for other functions using well-known ones, but later on we will explore definite integrals using these series (we will have no choice) and then bring in the wild card - the dilogarithm function.

Let's find the maclaurin series for

$$f(x) = \tan^{-1} x$$

174

From the previous page we have

$$\frac{1}{1+x} = 1 - x + x^2 - x^3 + \cdots = \sum_{n=0}^{\infty} (-1)^n x^n$$

from here we can change the parameter for x into x^2

$$\frac{1}{1+x^2} = 1 - (x^2) + (x^2)^2 - (x^2)^3 + \cdots = \sum_{n=0}^{\infty} (-1)^n (x^2)^n$$

$$\frac{1}{1+x^2} = 1 - x^2 + x^4 - x^6 + \cdots = \sum_{n=0}^{\infty} (-1)^n x^{2n}$$

and now integrate both sides. The left side we will recognize as arctangent!

$$\int \frac{1}{1+x^2} dx = \int (1 - x^2 + x^4 - x^6 + \cdots) dx = \sum_{n=0}^{\infty} \int (-1)^n x^{2n} dx$$

$$\tan^{-1}(x) = x - \frac{x^3}{3} + \frac{x^5}{5} - \frac{x^7}{7} + \cdots = \sum_{n=0}^{\infty} \frac{(-1)^n x^{2n+1}}{(2n+1)}$$

As you can see, playing with some of the well-known maclaurin series helps us find series for other functions without having to do too much work. Phew!

But why are we even talking about this?

what if I asked you to estimate:

$$e^{0.2} = ??$$

Sure you can plug it into the calculator, but a maclaurin series is used here!

Below are the first four non-zero terms of the series for e^x. Now watch what happens when we plug in x =0.2

$$e^x = 1 + x + \frac{x^2}{2!} + \frac{x^3}{3!}$$

$$e^{0.2} = 1 + (0.2) + \frac{(0.2)^2}{2!} + \frac{(0.2)^3}{3!} = 1.2213$$

compare that to the actual value and you'll see we are not *too* far off.

So what if you were asked to find the following definite integral? How would you go about it?

$$\int_0^1 \frac{1 - \cos x}{x^2} \, dx$$

certainly you can try some sort of integration technique. If you gave it a try you'd find the second term to be a pain to work with.

So instead we use a series!

176

Let's start out with the series for cosine and go from there! See if you can follow my steps.

$$\cos x = 1 - \frac{x^2}{2!} + \frac{x^4}{4!} - \frac{x^6}{6!} + \frac{x^8}{8!} - \frac{x^{10}}{10!} + \cdots$$

$$-\cos x = -1 + \frac{x^2}{2!} - \frac{x^4}{4!} + \frac{x^6}{6!} - \frac{x^8}{8!} + \frac{x^{10}}{10!} - \cdots$$

$$1 - \cos x = 1 - 1 + \frac{x^2}{2!} - \frac{x^4}{4!} + \frac{x^6}{6!} - \frac{x^8}{8!} + \frac{x^{10}}{10!} - \cdots$$

$$= \frac{x^2}{2!} - \frac{x^4}{4!} + \frac{x^6}{6!} - \frac{x^8}{8!} + \frac{x^{10}}{10!} - \cdots$$

$$\frac{1 - \cos x}{x^2} = \frac{x^2}{x^2\,2!} - \frac{x^4}{x^2\,4!} + \frac{x^6}{x^2\,6!} - \frac{x^8}{x^2\,8!} + \frac{x^{10}}{x^2\,10!} - \cdots$$

$$= \frac{1}{2!} - \frac{x^2}{4!} + \frac{x^4}{6!} - \frac{x^6}{8!} + \frac{x^8}{10!} - \cdots$$

$$\int_0^1 \frac{1 - \cos x}{x^2}\,dx = \int_0^1 \left(\frac{1}{2!} - \frac{x^2}{4!} + \frac{x^4}{6!} - \frac{x^6}{8!} + \frac{x^8}{10!} - \cdots \right) dx$$

$$= \left[\frac{1}{2!}x - \frac{x^3}{3 \cdot 4!} + \frac{x^5}{5 \cdot 6!} - \frac{x^7}{7 \cdot 8!} + \frac{x^9}{9 \cdot 10!} - \cdots \right]_0^1$$

$$= \frac{1}{2!} - \frac{1}{3 \cdot 4!} + \frac{1}{5 \cdot 6!} - \frac{1}{7 \cdot 8!} + \frac{1}{9 \cdot 10!}$$

$$\boxed{= 0.4863853}$$

Here I used the first five non-zero terms to get a more accurate answer.

Let's see if you can find the following definite integrals using series.

427) $\displaystyle\int_0^1 \frac{\sin x}{x}\, dx$

428) $\displaystyle\int_0^{1/2} \frac{\ln(1+x)}{x}\, dx$

429) $\displaystyle\int_0^{0.3} \frac{\tan^{-1} x}{x}\, dx$

430) $\displaystyle\int_{-1}^0 \frac{e^{3x} - 1}{10x}\, dx$

431) $\displaystyle\int_0^1 \frac{1 - e^{-x}}{x}\, dx$

432) $\displaystyle\int_{1/2}^{\pi/6} \frac{\sin(x) - 4}{x}\, dx$

178
The Dilogarithm Function

Well, nerds, this is the last section of integration techniques we'll encounter on this journey, in this first edition.

I say first edition because there will *always* be more to talk about. Whether it's due to demand or someone in the future finds another beautiful technique worth mentioning.

There are still plenty of integrals to discuss. Fresnel Integrals, Complex Integrals and Residue Theorem, the Gamma Function, and beyond. The point is, the learning will never stop.

But, instead of lamenting the end of our journey (for now!) look at this integral:

$$\text{Li}_2(z) = - \int_0^z \frac{\ln(1-u)}{u} \, du$$

this dilogarithm function proves useful in areas like physics and applications to Feynman integral calculations in quantum field theory. They are also tied to algebraic structures. The dilogarithm function is recognized as

$$\text{Li}_2(z) = \sum_{n=1}^{\infty} \frac{z^k}{k^2}$$

arriving at this solution requires - wouldn't you know it - the Maclaurin series for this natural log function!

We'll start off with the series given on page 169

$$\frac{1}{1-u} = \sum_{n=0}^{\infty} u^n$$

$$\frac{1}{1-u} = \sum_{n=1}^{\infty} u^{n-1}$$

notice that I changed the starting value of n from 0 to 1, so we adjust the power. Now we integrate both sides!

$$\int \frac{1}{1-u} \, du = \sum_{n=1}^{\infty} \int u^{n-1} \, du$$

$$-\ln(1-u) = \sum_{n=1}^{\infty} \frac{u^n}{n}$$

divide by u and integrate from 0 to z

$$-\frac{\ln(1-u)}{u} = \sum_{n=1}^{\infty} \frac{u^{n-1}}{n}$$

$$-\int_0^z \frac{\ln(1-u)}{u} \, du = \sum_{n=1}^{\infty} \int_0^z \frac{u^{n-1}}{n} \, du$$

180

Now we use first fundamental theorem of calculus and arrive at our solution!

$$= \sum_{n=1}^{\infty} \left[\frac{u^n}{n^2} \right]_0^z$$

$$= \sum_{n=1}^{\infty} \frac{z^n}{n^2}$$

I should note that z is complex value in this case!

It can be difficult to know all values for the dilogarithm function, but we know an important one.

$$\mathrm{Li}_2(1) = \sum_{n=1}^{\infty} \frac{1^k}{k^2} = \sum_{n=1}^{\infty} \frac{1}{k^2} = \frac{\pi^2}{6}$$

All this to say that if we encounter an integration problem that has these characteristics then we have no choice but to use the dilogarithm function!

Let's prove

$$\int_0^{\infty} \ln(1 - e^{-x}) \, dx = -\mathrm{Li}_2(1)$$

on the next page.

$$\frac{1}{1-x} = \sum_{n=1}^{\infty} x^{n-1}$$

$$\int \frac{1}{1-x}\, dx = \sum_{n=1}^{\infty} \int x^{n-1}\, dx$$

$$-\ln(1-x) = \sum_{n=1}^{\infty} \frac{x^n}{n}$$

$$-\ln(1-e^{-x}) = \sum_{n=1}^{\infty} \frac{e^{-xn}}{n}$$

$$-\int_{0}^{\infty} \ln(1-e^{-x})\, dx = \sum_{n=1}^{\infty} \int \frac{e^{-xn}}{n}\, dx$$

$$-\int_{0}^{\infty} \ln(1-e^{-x})\, dx = \sum_{n=1}^{\infty} \left[-\frac{e^{-xn}}{n^2} \right]_{0}^{\infty}$$

$$-\int_{0}^{\infty} \ln(1-e^{-x})\, dx = \sum_{n=1}^{\infty} \lim_{b \to \infty} -\frac{e^{-bn}}{n^2} + \frac{1}{n^2}$$

Since $\lim\limits_{b \to \infty} -\dfrac{e^{-bn}}{n^2} = 0$

$$-\int_{0}^{\infty} \ln(1-e^{-x})\, dx = \sum_{n=1}^{\infty} \frac{1}{n^2}$$

$$\int_0^\infty \ln\left(1 - e^{-x}\right) dx = -\sum_{n=1}^\infty \frac{1}{n^2}$$

$$\boxed{= -\frac{\pi^2}{6}}$$

Techniques like this seem sooooo simple. And it sorta is, since alot of the groundwork has been laid out for us. Someone has a perfect formula for the dilogarithm function and how it relates to the series and a natural log integral, but it's our job to cut away at problems and arrive at the correct solution!

At this level problems take alot of cleverness and due diligence. Minor mistakes can crumble your work and perhaps more difficult to spot. Arriving at the solution is oh-so satisfying, however!

Let's see how clever you can be with the following:

433) $\displaystyle\int_0^\infty \ln\left(\frac{e^x - 1}{e^x + 1}\right) dx$

434) $\displaystyle\int_0^\infty \ln(1 + e^{-x}) \, dx$

435) $\displaystyle\int_0^\infty \ln(2^x - 1) - \ln(2^x) \, dx$

436) Find $\mathrm{Li}_2(i)$ where $i = \sqrt{-1}$

437) Prove $\dfrac{d}{dz}\left[\mathrm{Li}_2\left(\dfrac{1}{z}\right)\right] = -\dfrac{\ln\left(\dfrac{z-1}{z}\right)}{z}$

438) $\displaystyle\int \dfrac{\tan^{-1}(3x)}{x}\, dx$

Real part

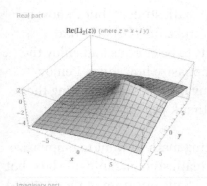

Real part

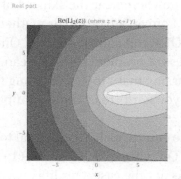

Imaginary part

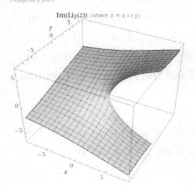

Imaginary part

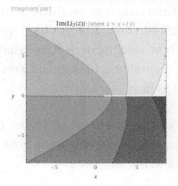

184

The Work Never Ends

Think about all the great mathematicians that came before us. Euler, Euclid, Newton, Leibniz, Taylor, DeMoivre's, Descartes, Fermat, the list goes on and on (no, I'm not dissing those I left out!)

They set the rules for us to follow and to this day mathematical minds are creating and discovering new ideas, techniques, and solutions to complex problems.

This is all because we continue to explore the curious world of mathematics. We continue to share ideas whether it's through videos, social settings, or books like this. The field of mathematics needs you! What use will you make of these integrals?

I am nowhere near the skill set of a PhD math student, but I know my lane. I am an advocate for mathematics and a firm believer that ANYONE can contribute to it. One needs to be curious and enjoy the process of problem solving. I cannot emphasize how much I enjoy discussing math topics, learning from my mistakes, and sharing them with you nerds to critique or unpack. I thank you for that.

At the end curiosity is essential to learning math - I want to emphasize that. Often times schools want to fit all math students into one box, but that is not the case. Our life experiences shape our train of thought and sometimes the way we think of math concepts way differently than your classmates. That is okay! As long as we are following the same rules how we approach it, or what route we take, is totally up to you. Solve it whatever fucken way you want. Take the scenic route, cause THAT is how you learn.

Your journey does not end here, this is just a small portion of it. Go learn and explore, cause the next great mathematical discovery is among us!

theveganmathguy.com

Answer Key

This particular book is so special to me. Not only did I try my absolute best to write and explain techniques, but the answer key has been solved by the collaborative effort of some math nerds! As I wrote the book I called out to my followers to help find and write solutions to the integrals, and 4 individuals stepped up. I am forever grateful to the following people:

- Károly Bendegúz Bók (@bkbendeguz7)
- Edward Hawkins
- Jake Rogers (@jakeersan / @satmathtips)
- Jenny "feli" Y

1.
$$\frac{x^3}{3} + \frac{3x^2}{2} + c$$

2.
$$4x^{3/2} - \frac{2}{x^2} + C$$

3.
$$\frac{21}{4}x^{\frac{4}{3}} - 3x^2 + 11x + c$$

4.
$$4x + \frac{8}{x} + \frac{x^4}{4} + c$$

5.
$$\frac{4}{5}x^{\frac{5}{4}} + 4x^{\frac{3}{2}} + \frac{7}{x} + c$$

6.
$$\frac{5}{12}x^{12} - (x+1)^3 + c$$

7.
$$\frac{21}{4}x^{\frac{8}{7}} + \frac{8}{x} + 8x + c$$

8.
$$2x^3 - 2x^2 + x - \frac{11}{12}x^{\frac{12}{11}} + c$$

9.
$$\frac{x^7}{7} + \frac{9}{5}x^5 + 9x^3 + 27x + c$$

10.
$$\frac{12}{7}x^{\frac{7}{2}} - \frac{8}{5}x^{\frac{5}{2}} + 200x^{\frac{1}{2}} + c$$

11.
$$8x^2 - \frac{48}{11}x^{\frac{11}{6}} + \frac{3}{5}x^{\frac{5}{3}} + c$$

12.
$$\frac{4}{9}x^{\frac{9}{4}} - 2x^3 + c$$

13.
$$\frac{x^5}{5} + \frac{2}{3}x^3 + x + \frac{2}{5}x^{\frac{5}{2}} + \frac{2}{3}x^{\frac{3}{2}} + c$$

14.
$$\frac{x^{13}}{13} + \frac{x^9}{3} + \frac{3}{5}x^5 + x + c$$

15.
$$\frac{18}{7}x^{\frac{7}{3}} + \frac{2}{3}x^{-\frac{3}{2}} + c$$

16.
$$\frac{15}{4}x^4 - 7\left(\frac{x^3}{3} + \frac{x^2}{2} - x\right) + c$$

17.
$$\frac{14}{3}x^{\frac{3}{2}} + \frac{52}{5}x^{\frac{5}{4}} + \frac{1}{x} + c$$

18.
$$\frac{x^8}{8} + \frac{8}{9}x^{\frac{9}{2}} + 4x + c$$

19.
$$\frac{2}{3}x^6 + \frac{12}{5}x^{\frac{5}{2}} - \frac{8}{3}x^{\frac{9}{4}} - \frac{6}{x^2} + c$$

20.
$$-\cos x + \sin x + c$$

21. $\tan(x) + x + C$

22. $e^x + \dfrac{6^x}{\ln(6)} + C$

23. $6\sin(x) + \dfrac{10x^{\frac{3}{2}}}{3} + C$

24. $6\ln|x| - \sin(x) + e^x + C$

25. $4(\tan(x) + 2x) + C$

26. $7e^x + \dfrac{7^x}{\ln(7)} + C$

27. $2\left(\sqrt{x} - 6\cos(x)\right) + C$

28. $-\cot(x) + \sec(x) + C$

29. $\dfrac{2^x}{\ln 2} + 5\ln|x| + 10x + \dfrac{5x^2}{2} + C$

30. $\sec(x) + C$

31. $e^x - 3x + C$

32. $3\ln|x| + e^x + \dfrac{4 \cdot 3^x}{\ln(3)} + C$

33. $6\ln|x| + C$

34. $-\cot(x) - \tan(x) + C$

35. $e^x - \dfrac{1}{x} + 2\ln|x| + x + C$

36) $\dfrac{8^x}{3\ln(2)} + \cot(x) + \tan(x) + C$

37) $\dfrac{2^{2x-1}}{\ln(2)} + \sec(x) + \cos(x) + C$

38) $4\ln|x| + \dfrac{4}{x} + \dfrac{2^{2x-1}}{\ln(2)} + C$

39) $\tan(x) + C$

40) $\dfrac{1}{18}(3x+8)^6 + C$

41) $4\ln|x+8| + C$

42) $\dfrac{(3x^2+1)^5}{30} + 4\ln|x^2-3| + 2C$

43) $-\dfrac{1}{2}\cos(2x) + \dfrac{1}{6}e^{6x-1} + C$

44) $\ln|x^3 + 6x + 8| + C$

45) $\dfrac{1}{2}e^{x^2+8} - 2\ln|x+3| + C$

46) $-\dfrac{1}{2}\ln|\sin(2x)| - \ln|\cos(2x)|$

47) $\frac{1}{18}\left(6x^2 - 1\right)^{\frac{3}{2}} + C$

48) $2\ln|x^2 + 6x - 1| + C$

49) $\sec(2x) + C$

50) $\ln|\tan(x)| - \frac{\sin^2(x)}{2} + C$

51) $e^{2+x^2+x} + C$

52) $\frac{1}{2}e^{2x} + x + C$

53) $\frac{2^{2\sin(x)-1}}{\ln(2)} - \frac{3}{4}\ln|4x + 5| + C$

54) $2\ln|\sqrt{x} + 1| + 2\sqrt{x} + C$

55) $-\frac{1}{3}\ln|\cos(\sqrt{x})| + C$

56) $\frac{1}{3}\left(x^2 + 1\right)^6 + C$

57) $\frac{2^{10x-13}}{5\ln 2} - \frac{1}{6(e^x + 1)^6} + \frac{3}{4}\ln|4x + 1| + C$

58) $\frac{8}{45}\left(x^3 + 1\right)^{3/2}\left(3x^3 - 2\right) + \text{constant}$

59) $\frac{1}{15}\left(3x^2 + 2\right)\left(x^2 - 1\right)^{\frac{3}{2}} + C$

60) $3\left(\frac{3}{7}\left(x^2 + 1\right)^{\frac{7}{3}} - \frac{3}{4}\left(x^2 + 1\right)^{\frac{4}{3}}\right) + C$

61) $2\left(\frac{2}{7}\left(\sqrt{x} + 6\right)^{\frac{7}{2}} - \frac{24}{5}\left(\sqrt{x} + 6\right)^{\frac{5}{2}} + 24\left(\sqrt{x} + 6\right)^{\frac{3}{2}}\right) + C$

62) $\frac{1}{3}\left(x^2 + 1\right)^{\frac{3}{2}} - \sqrt{x^2 + 1} + C$

63) $\quad \frac{1}{15} \sqrt{x^2 - 1} \left(3x^4 + 4x^2 + 8\right)$ + constant

64) $\quad \frac{1}{21} \left(x^2 + 10\right)^{3/2} \left(3x^4 - 24x^2 + 160\right)$ + constant

65) $\quad \frac{15}{7} \left(x^2 - 6\right)\left(x^2 + 8\right)^{4/3}$ + constant

66) $\quad 2\left(x^2 - 1\right)^{\frac{3}{2}} + 5\sqrt{x^2 - 1} + C$

67. $\quad \dfrac{\left(x^2 + 2\right)^{\frac{3}{2}} \left(15x^4 - 3x^2 + 4\right)}{105}$

68. $\quad \dfrac{\arctan\left(\frac{2x+4}{10}\right)}{5}$

69. $\quad \dfrac{\ln\left(|3x + 5|\right) - \ln\left(|x - 1|\right)}{24} + C$

70. $\quad \ln\left|\sin(x) + 1\right| + C$

71 $\quad \dfrac{\ln\left(|\tan(4x) + \sec(4x)|\right) + \sec(4x)\tan(4x)}{8} + C$

72 $\quad \dfrac{\sin(8x+2)}{16} - \dfrac{\sin(6x+2)}{12} + x + C$

73 $\quad \dfrac{\arctan\left(\frac{x^2+1}{6}\right)}{4} + C$

74 $\quad -\dfrac{\sin(6x) - 6x}{12} + C$

75 $\quad \dfrac{\ln\left(|x - \sqrt{13}|\right) - \ln\left(|x + \sqrt{13}|\right)}{2\sqrt{13}} + C$

76 $\quad \dfrac{3\arctan\left(\frac{4x+15}{\sqrt{10}}\right)}{2\sqrt{10}} + C$

77 $\quad \ln\left(|\tan(x) + \sec(x)|\right) + C$

78 $\quad \dfrac{-\sin(2x) - 2x}{4} + C$

79 $\cos(x)\sin(x) + C$

80 $\dfrac{\arctan\left(\frac{x+5}{6}\right)}{6} + C$

81 $\dfrac{\ln\left(\left|x-2\sqrt{6}+5\right|\right) - \ln\left(\left|x+2\sqrt{6}+5\right|\right)}{4\sqrt{6}} + C$

82 $\dfrac{\ln\left(\left|2x-\sqrt{14}+6\right|\right) - \ln\left(\left|2x+\sqrt{14}+6\right|\right)}{2\sqrt{14}} + C$

83 $\dfrac{\arctan\left(\frac{6x+12}{4\sqrt{51}}\right)}{2\sqrt{51}} + C$

84 $\dfrac{2\arctan\left(\frac{2e^x+1}{\sqrt{23}}\right)}{\sqrt{23}} + C$

85 $-\dfrac{1}{2\sqrt{37}}\left(\ln\left|\dfrac{4\sin(x)+5}{\sqrt{37}}+1\right| - \ln\left|\dfrac{4\sin(x)+5}{\sqrt{37}}-1\right|\right) + C$

86 $\dfrac{2\arctan\left(\frac{6x-5}{\sqrt{95}}\right)}{\sqrt{95}} + C$

87 $\dfrac{x\sqrt{x^2-16}}{2} - 8\ln\left(\left|\sqrt{x^2-16}+x\right|\right) + C$

88 $\dfrac{49\ln\left(\left|\sqrt{x^2+49}+x\right|\right) + x\sqrt{x^2+49}}{2} + C$

$= \dfrac{x\sqrt{x^2+49} + 49\,\mathrm{arsinh}\left(\frac{x}{7}\right)}{2} + C$

89 $\dfrac{x\sqrt{25-x^2} + 25\arcsin\left(\frac{x}{5}\right)}{2} + C$

90 $\arcsin\left(\frac{x}{2}\right) + C$

91 $\ln(\sqrt{x^2+10}+x|) + C$
$= \mathrm{arsinh}\left(\frac{x}{\sqrt{10}}\right) + C$

92 $\ln(|\sqrt{x^2 - 81} + x|) + C$

93 $\ln(|\sqrt{x^2 + 6x - 72} + x + 3|) + C$

94 $\ln(|x + \sqrt{(x-3)^2 + 4} - 3|) + C$
$= \text{arsinh}\left(\frac{x-3}{2}\right) + C$

95 $\arcsin\left(\frac{x-1}{3}\right) + C$

96 $\frac{\ln(|\sqrt{(3x+1)^2 - 5} + 3x + 1|)}{3} + C$

97 $\ln\left(|\sqrt{\sin^2(x) + 1} + \sin(x)|\right) + C$
$= \text{arsinh}(\sin(x)) + C$

98 $\frac{\sqrt{9x^2 - 25}}{25x} + C$

99 $\frac{\arctan\left(\frac{x+1}{2}\right)}{2} + C$

100 $-\frac{\arctan\left(\frac{\sqrt{16-x^4}}{x^2}\right)}{2} + C$
$= \frac{\arcsin\left(\frac{x^2}{4}\right)}{2} + C$

101 $\ln(|\sqrt{(x+3)^2 - 25} + x + 3|) + C$

102 $\arcsin\left(\frac{x+1}{2}\right) + C$

103 $\frac{\ln(|\sqrt{x^2+1} - 1|)\ln(|\sqrt{x^2+1} + 1|)}{2} + \sqrt{x^2 + 1} + C$
$= \sqrt{x^2 + 1} - \text{arsinh}\left(\frac{1}{|x|}\right) + C$

104 $\sqrt{x^2 + x + 1} - \frac{\ln(|2(\sqrt{x^2+x+1} + x) + 1|)}{2} + C$
$= \sqrt{x^2 + x + 1} - \frac{\text{arsinh}\left(\frac{2x+1}{\sqrt{3}}\right)}{2} + C$

105 $\ln(|x + \sqrt{(x-3)^2 + 4} - 3|) + C$
$= \text{arsinh}\left(\frac{x-3}{2}\right) + C$

106 $\frac{5\ln(\sqrt{2x^4 + 3} + \sqrt{2}x^2)}{2^{\frac{3}{2}}} + C$

107 $\frac{4\ln\left(\left|\frac{\sqrt{7-(x+1)^2} - \sqrt{7}}{x+1} + \sqrt{7} - \sqrt{6}\right|\right) - 4\ln\left(\left|\frac{\sqrt{7-(x+1)^2} - \sqrt{7}}{x+1} + \sqrt{7} + \sqrt{6}\right|\right)}{\sqrt{6}} + C$
$= -\frac{4\ln\left(\left|\frac{\sqrt{6}\sqrt{-x^2-2x+6}}{|x|} + \frac{6}{|x|} - 1\right|\right)}{\sqrt{6}} + C$

108 $\ln(|x - 6|) - \ln(|x + 2|) + C$

109 $\frac{1}{3}\ln|x| - \frac{1}{3}\ln|x + 3| + C$

110 $-\frac{\ln(|x-3||x+4|)+3\ln(|x+4|)-3\ln(|x-3|)}{2}+C$

$= \ln(|x-3|) - 2\ln(|x+4|) + C$

111 $\frac{9\ln(|x+1||x+4|)+19\ln(|x+4|)-19\ln(|x+1|)}{6}+C$

$= \frac{14\ln(|x+4|)-5\ln(|x+1|)}{3}+C$

112) $6\ln(x+4) - \ln(7-x) + C$

113) $\frac{11}{3}\ln(2-x) + \frac{4}{3}\ln(x+1) + C$

114) $\frac{1}{2}\ln(1-x^2) - \ln(x) + C$

115) $\ln(x^2+4) + \ln(x-3) + C$

116) $\ln(x-3) - \tan^{-1}(\frac{x}{2}) + C$

117) $-\ln(1-x) - 2\tan^{-1}(\frac{3}{x}) + C$

118) $\ln(x-1) + 2\tan^{-1}(x) + C$

119) $4\ln(x+1) + \frac{1}{9}\tan^{-1}(\frac{x}{3}) + C$

120) $\frac{1}{3}\left(-\ln(3-x) + \ln(x+3) + 18\tan^{-1}(x)\right)+C$

121) $\frac{1}{6}\left(-\ln(x^2-x+1) + 2\ln(x+1) + 2\sqrt{3}\tan^{-1}(\frac{2x-1}{\sqrt{3}})\right)+C$

122) $\frac{1}{243}\left(-\ln(x^2-3x+9) + 2\ln(x+3) + 2\sqrt{3}\tan^{-1}(\frac{2x-3}{3\sqrt{3}})\right)+C$

123) $2\left(-\frac{1}{6}\ln(x^2+x+1) + \frac{1}{x} + \frac{1}{3}\ln(1-x) + \frac{\tan^{-1}(\frac{2x-3}{\sqrt{3}})}{\sqrt{3}}\right)+C$

124) $2\left(\dfrac{3}{x+2} + \ln(x+2)\right) + C$

125) $-\dfrac{1}{2}\ln(x^2+1) + \dfrac{1}{x-1} + \ln(x-1) + \tan^{-1}(x) + C$

126) $\dfrac{3-4x}{2(x-1)^2} + \ln(x-1) + C$

127) $\dfrac{1}{x+1} + \dfrac{1}{2}\sqrt{5}\ln(\sqrt{5}-x) + 4\ln(x+1) - \dfrac{1}{2}\sqrt{5}\ln(x+\sqrt{5}) + C$

128) $\dfrac{3\tan^{-1}\left(\frac{x+3}{\sqrt{91}}\right)}{\sqrt{91}} - 4\ln(x+5) + C$

129) $\dfrac{1}{2}\ln(x^2+10) + \dfrac{5}{x+1} + 7\ln(x+1) + C$

130) $\dfrac{18x^2+33x+19}{6(x+1)^3} + \ln(x+1) + C$

131) $\dfrac{x^2}{2} + x - 3\ln(x-8) + C$

132) $\dfrac{1}{2}x(x+10) + \ln(x-6) + C$

133.

$$\frac{2}{3}x^3 + 2x^2 - 3x + \frac{1}{2}\ln|2x - 1| + c$$

134.

$$x + \frac{9}{5}\ln|x + 4| + \frac{6}{5}\ln|x - 1| + c$$

135.

$$x + 2\ln|x + 1| - 5\ln|x + 2| + c$$

136.

$$\frac{3}{2}x^2 - 5\ln|x + 1| + 7\ln|x - 1| + c$$

137.

$$2x + 2\ln|x| - \ln|x - 5| + c$$

138.

$$\frac{x^3}{3} + x + \frac{1}{2}\ln\left|\frac{x - 1}{x + 1}\right| + c$$

139.

$$\frac{x^4}{4} - \frac{4}{3}x^3 - \frac{3}{2}x^2 + 6x + 22\ln|x + 2| +$$

140.

$$\frac{2}{3}x^3 + 4x^2 + 16x + 37\ln|x - 2| + c$$

141.

$$\frac{x^3}{3} - \frac{x^2}{2} + x - \ln|x^2 + 3x - 2| + c$$

142.

$$\frac{x^3}{3} - \frac{5}{2}x^2 + 6x + \ln|x + 1| + c$$

143.

$$\frac{3}{2}x^2 - x + \frac{1}{2}\arctan\left(\frac{x}{2}\right) + c$$

196

144. $\dfrac{11\ln|5x-4|}{5} + \dfrac{2x^3 - 3x^2 + 24x}{6} + C$

145. $8\arctan(x) - \dfrac{\ln\left|x^2+1\right|}{2} + 2x^3 - x^2 - 6x + C$

146. $\dfrac{3x^4 + 10x^2 + 16x}{4} - \ln\left|x^2 - 2\right| + C$

147. $\dfrac{95\ln|2x-1|}{64} + \dfrac{48x^5 + 90x^4 - 20x^3 - 15x^2 - 15x}{480} + C$

148. $x - 2\ln|x+1| + 4\ln\left(x^2 - x + 1\right) - \dfrac{4\arctan\left(\frac{2x-1}{\sqrt{3}}\right)}{\sqrt{3}} + C$

149. $\dfrac{181\ln|5x-1|}{780} + \dfrac{209\ln|x+5|}{104} - \dfrac{25\ln|x+1|}{24} + C$

150. $6\ln|x-1| - \ln|x+3| + 2\ln|x+2| + C$

151. $3\arctan\left(\dfrac{x}{2}\right) - \dfrac{\ln|2x-1|}{2} + C$

152. $\dfrac{\ln|x+3| - \ln|x-3|}{2} + \arctan\left(\dfrac{x}{2}\right) + 5\ln|x-1| + C$

153. $\dfrac{\ln\left|2x - 3\sqrt{13} + 11\right| - \ln\left|2x + 3\sqrt{13} + 11\right|}{\sqrt{13}} + \ln|x-2| + C$

154. $-\dfrac{4\ln|5x-1|}{5} + 6\ln|x| - \dfrac{1}{x+1} + C$

155. $\dfrac{\ln|3x-1|}{3} + \ln|x+3| + \ln|x+1| + C$

156. $\dfrac{\sqrt{6}\left(\ln\left|2x+\sqrt{6}\right| - \ln\left|2x-\sqrt{6}\right|\right) - 2\ln\left|x^2-2\right|}{4} + 4\ln|x-2| + C$

157.

$$\ln|3x^2 - 9x + 1| - \dfrac{5}{\sqrt{69}}\ln|6x - 9 + \sqrt{69}| + \dfrac{5}{\sqrt{69}}\ln|-6x + \sqrt{69} + 9| + c$$

158.

$$\dfrac{1}{4}\ln|2x^2 + x + 8| - \dfrac{13}{6\sqrt{7}}\arctan\left(\dfrac{4x+1}{3\sqrt{7}}\right) + c$$

159.

$$\dfrac{5}{2}\ln|x^2 + x + 4| - \sqrt{\dfrac{3}{5}}\arctan\left(\dfrac{2x+1}{\sqrt{15}}\right) + c$$

160.
$$\frac{7}{6}\ln|3x^2 + 5x + 4| - \frac{\sqrt{23}}{3}\arctan\left(\frac{6x+5}{\sqrt{23}}\right) + c$$

161.
$$\frac{3}{2}\ln|x^2 + 5x + 15| - 3\sqrt{\frac{7}{5}}\arctan\left(\frac{2x+5}{\sqrt{35}}\right) + c$$

162.
$$3\ln|x^2 + 10x + 26| - 31\arctan(x+5) + c$$

163.
$$\ln|x^2 - 3x + 100| + \frac{16}{\sqrt{391}}\arctan\left(\frac{2x-3}{\sqrt{391}}\right) + c$$

164.
$$\frac{76 + 13\sqrt{95}}{190}\ln|-5x + \sqrt{95} + 10| + \frac{76 - 13\sqrt{95}}{190}\ln|5x + \sqrt{95} - 10| + c$$

165.
$$\frac{41 - 115\sqrt{41}}{656}\ln|-16x + \sqrt{41} - 3| + \frac{41 + 115\sqrt{41}}{656}\ln|16x + \sqrt{41} + 3| + c$$

166.
$$\frac{1}{4}\ln|6x^2 - x + 2| + \frac{401}{2\sqrt{47}}\arctan\left(\frac{12x-1}{\sqrt{47}}\right) + c$$

167) $x\sin(x) + \cos(x) + C$

168) $\dfrac{1}{16}x^4(4\ln(x) - 1)) + C$

169) $\cos(x) - \cos(x)\ln(\cos(x)) + C$

170) $-\dfrac{1}{5}(4 - x^2)^{\frac{3}{2}}(3x^2 + 8) + C$

171) $2x\sin(x) - (x^2 - 2)\cos(x) + C$

172) $-\dfrac{1}{5}(9-x^2)^{\frac{3}{2}}(x^2+6)+C$

173) $-\dfrac{1}{2\ln^2(x)}+C$

174) $\dfrac{1}{9}\Big(\cos(1-6x)-6x\sin(1-6x)\Big)+C$

175) $\dfrac{1}{25}(1+4e^5)\approx 23.786$

176) $\sqrt{1-x^2}+\sin^{-1}x+C$

177) $x\Big(\ln(x)-1\Big)+C$

178) $x\tan^{-1}x-\dfrac{1}{2}\ln(x^2+1)+C$

179) $x\cos^{-1}x-\sqrt{1-x^2}+C$

180) $x\Big(\ln(x^2+1)-2\Big)+2\tan^{-1}x+C$

181) $x\Big(\ln(x^2+2x-3)-2\Big)-\ln(1-x)+3\ln(x+3)+C$

182) $\dfrac{1}{6}\Big((6x+1)\ln(3x^2+x+1)-12x+2\sqrt{11}\tan^{-1}\dfrac{6x+1}{\sqrt{11}}\Big)+C$

183) $x\left(\ln(x^2-1)-2\right)+2\tanh^{-1}x+C$

184) $x\left(\ln(x^2+4)-2\right)+4\tan^{-1}\dfrac{x}{2}+C$

185) $x\left(\ln(x^2+a^2)-2\right)+2a\tanh^{-1}\dfrac{x}{a}+C, a\epsilon\mathbb{Z}$

186) $x\left(\ln(x^2-6x+1)-2\right)+(2\sqrt{2}-3)\ln(-x-2\sqrt{2}+3)-(3+2\sqrt{2})$
$\ln(-x+2\sqrt{2}+3)+C$

187) $x\left(\ln(x^2-10x+1)-2\right)+(2\sqrt{6}-5)\ln(-x-2\sqrt{6}+5)$
$-(5+2\sqrt{6})\ln(-x+2\sqrt{6}+5)+C$

188) $\dfrac{1}{2}e^x\left(\sin(x)+\cos(x)\right)+C$

189) $-\dfrac{1}{5}e^{2x}\left(\cos(x)-2\sin(x)\right)+C$

190) $\dfrac{1}{20}e^{2x}\left(3\sin(6x+2)\cos(6x+2)\right)+C$

191) $-\cos(x)+\dfrac{1}{3}\cos^3(x)+C$

193. $2\left(\tan(x^4)\sec(x^4) - \log\left(\cos\left(\frac{x^4}{2}\right) - \sin\left(\frac{x^4}{2}\right)\right) + \log\left(\sin\left(\frac{x^4}{2}\right) + \cos\left(\frac{x^4}{2}\right)\right)\right) + \text{constant}$

194. $\dfrac{x^4\sin(2x)}{2} - \dfrac{3x^2\sin(2x)}{2} + \dfrac{3\sin(2x)}{4} + x^3\cos(2x) - \dfrac{3x\cos(2x)}{2} + C$

195. $\dfrac{e^{2x}\left(2x^4 - 4x^3 + 6x^2 - 6x + 3\right)}{4} + C$

196. $\dfrac{x\sin(2x)}{2} - \dfrac{(x^2+1)\cos(2x)}{2} + \dfrac{\cos(2x)}{4} + C$

197. $\left(4x^3 - 6x\right)\sin(2x) + \left(6x^2 - 3\right)\cos(2x) + C$

198. $e^x\left(8x^3 - 12x^2 + 30x - 29\right) + C$

199) $10x\left(x^8 - 72x^6 + 3024x^4 - 60480x^2 + 362880\right)\sin(x) - \left(x^{10} - 90x^8 + 5040x^6 - 151200x^4 + 1814400x^2 - 3628800\right)\cos(x) + \text{constant}$

200. $4x\sin(2x+1) + \left(2 - 4x^2\right)\cos(2x+1) + C$

201) $\dfrac{1}{126}\sin^7\left(7\cos(2x) + 11\right) + C$

202) $-\dfrac{1}{840}\cos^3(x)\left(-108\cos(2x) + 15\cos(4x) + 157\right) + C$

203) $\dfrac{-72\cos(2x) + 12\cos(4x) + 8\cos(6x) - 3\cos(8x)}{3072} + C$

204) $\dfrac{1}{120120}\sin^5(x)\left(171400\cos(2x) + 61400\cos(4x) + 3(4200\cos[6x])\right.$
$\left. + 385\cos(8x) + 45923\right) + C$

205) $\dfrac{1}{30}\cos^3(x)\left(3\cos(2x) - 7\right) + C$

206) $\dfrac{1}{32}\left(12x - 8\sin(2x) + \sin(4x)\right) + C$

207) $\dfrac{1}{8}\left(4x + \sin(4x)\right)+C$

208) $\dfrac{1}{3}\sec(x)\left(\sec^2(x) - 3\right)+C$

209) $\dfrac{1}{4}\left(\sec^4(x) - 4\sec^2(x) - 4\ln(\cos[x])\right)+C$

210) $\dfrac{1}{96}\left(-4\cos(2x) + 3\cos(4x) + 5\right)+ \sec^8(x) + C$

211) $\dfrac{1}{15}\left(6\cos(2x) + \cos(4x) + 8\right)+ \tan(x)\sec^4(x) + C$

212) $\dfrac{1}{3}\left(\cos(2x) + 2\right)+ \tan(x)\sec^2(x) + C$

213) $\dfrac{1}{12}\left(-3\tan^{-1}(\tan 1 - 4x)\right)- \tan^3(1 - 4x) + 3\tan(1 - 4x) + C$

214) $\dfrac{1}{15}\sec(5x)\left(\sec^2(5x) - 3\right)+C$

215) $\dfrac{3}{2}\sec^2(\dfrac{x}{3}) + C$

216) $\dfrac{1}{16}\left(\dfrac{1}{2}(11\sin[x]+3\sin[3x])\sec^4(x)-6\ln(\cos[\frac{x}{2}]-\sin[\frac{x}{2}])+6\ln(\sin[\frac{x}{2}]\right.$
$\left.+ \cos[\frac{x}{2}])\right)+C$

217. $2\displaystyle\int_0^{\frac{\pi}{2}} \cos^3(x)\,dx = 2\int_0^{\frac{\pi}{2}} \cos^2(x)\cdot\cos(x)\,dx$

$$= 2\int_0^{\frac{\pi}{2}} (1-\sin^2(x))\cdot\cos(x)\,dx$$

$$= 2\int_0^1 (1-u^2)\,du$$

$$= 2\left[u - \frac{u^3}{3}\right]_0^1$$

$$= 2\left(1 - \frac{1}{3}\right) - 2\,(0-0)$$

$$= \frac{4}{3}.$$

218. $4\displaystyle\int_0^{\frac{\pi}{2}} \sin^2(x)\,dx = 4\int_0^{\frac{\pi}{2}} \frac{1-\cos(2x)}{2}\,dx$

$$= 2\int_0^{\frac{\pi}{2}} (1-\cos(2x))\,dx$$

$$= 2\left[x - \frac{\sin(2x)}{2}\right]_0^{\frac{\pi}{2}}$$

$$= 2\left[\frac{\pi}{2} - \frac{\sin(\pi)}{2} - \left(0 - \frac{\sin(0)}{2}\right)\right]$$

$$= 2\left[\frac{\pi}{2} - \frac{0}{2} - 0\right]$$

$$= \pi$$

219. $\displaystyle\int_0^{\frac{\pi}{2}} \cos^7(x)\,dx = \frac{1}{7}\cos^6(x)\sin(x) + \frac{6}{7}\int_0^{\frac{\pi}{2}} \cos^5(x)\,dx$

$$= \frac{1}{7}\cos^6(x)\sin(x) + \frac{6}{7}\left(\frac{1}{5}\cos^4(x)\sin(x) + \frac{4}{5}\int_0^{\frac{\pi}{2}} \cos^3(x)\,dx\right)$$

$$= \frac{1}{7}\cos^6(x)\sin(x) + \frac{6}{35}\cos^4(x)\sin(x) + \frac{24}{35}\int_0^{\frac{\pi}{2}} \cos^3(x)\,dx$$

$$= \frac{1}{7}\cos^6(x)\sin(x) + \frac{6}{35}\cos^4(x)\sin(x) + \frac{24}{35}\left(\frac{1}{3}\cos^2(x)\sin(x) + \frac{2}{3}\int_0^{\frac{\pi}{2}} \cos(x)\,dx\right)$$

$$= \frac{1}{7}\cos^6(x)\sin(x) + \frac{6}{35}\cos^4(x)\sin(x) + \frac{8}{35}\cos^2(x)\sin(x) + \frac{16}{35}$$

220. $-\dfrac{\cos{(5x)}}{10} - \dfrac{\cos{(x)}}{2} + C$

221. $\dfrac{\sin{(7x)}}{14} + \dfrac{\sin{(x)}}{2} + C$

222. $\dfrac{\sin{(2x)}}{4} - \dfrac{\sin{(4x)}}{8} + C$

223. $\dfrac{\cos{(7x)}}{14} + \dfrac{\cos{(x)}}{2} + C$

224. $\ln\left|\tan\left(\dfrac{x}{2}\right) + 1\right| + C$

225. $\dfrac{2\arctan\left(\dfrac{\tan\left(\frac{x}{2}\right)}{\sqrt{3}}\right)}{\sqrt{3}} + C$

226. $2\arctan\left(3\tan\left(\dfrac{x}{2}\right)\right) + C$

227. $\dfrac{12\arctan\left(\dfrac{3\tan\left(\frac{x}{2}\right) - 2}{\sqrt{5}}\right)}{\sqrt{5}} + C$

228. $\dfrac{\tan{(x)}}{2} + C$

229. $x - \tan\left(\dfrac{x}{2}\right) + C$

230. $\dfrac{\ln\left|\tan\frac{x}{2} + \sqrt{2} - 1\right| - \ln\left|\tan\left(\frac{x}{2}\right) - \sqrt{2} - 1\right|}{\sqrt{2}} + C$

231. $x + \dfrac{2}{\tan\left(\frac{x}{2}\right) + 1} + C$

232. $\dfrac{\arctan\left(\sqrt{2}\tan(x)+1\right)+\arctan\left(\sqrt{2}\tan(x)-1\right)}{\sqrt{2}}+C$

233. $\dfrac{\ln\left|3\tan\left(\frac{x}{2}\right)+2\right|-\ln\left|2\tan\left(\frac{x}{2}\right)-3\right|}{13}+C$

234. $\dfrac{\ln\left|5\tan\left(\frac{x}{2}\right)+2\right|}{5}+C$

235. $\dfrac{4\arctan\left(\frac{\sqrt{3}\tan\left(\frac{x}{2}\right)}{\sqrt{5}}\right)}{\sqrt{15}}+C$

236) $\sqrt{2}\tan^{-1}(\sqrt{2}\sqrt{x+1})+C$

237) $\dfrac{\tan^{-1}\left(\frac{x-3}{\sqrt{2}\sqrt{x^2-3}}\right)}{\sqrt{2}}+C$

238) $-\dfrac{\sqrt{(x+1)^2+1}}{x+1}+C$

239) $\dfrac{2\tanh^{-1}\left(\frac{\sqrt{4x+1}}{\sqrt{7-2\sqrt{5}}}\right)}{\sqrt{5(7-2\sqrt{5})}}-\dfrac{2\tanh^{-1}\left(\frac{\sqrt{4x+1}}{\sqrt{7+2\sqrt{5}}}\right)}{\sqrt{5(7+2\sqrt{5})}}+C$

240) $-\dfrac{\tanh^{-1}\left(\frac{7x+2}{2\sqrt{15}\sqrt{x^2-x+3}}\right)}{\sqrt{15}}+C$

241) $\dfrac{1}{2}\tan^{-1}\left(\dfrac{2x}{\sqrt{x^2+5}}\right)+C$

242) $\dfrac{1}{\sqrt{2}} \tan^{-1}(\dfrac{\sqrt{x^2 + \frac{1}{x^2}}}{\sqrt{2}}) + C$

243) $\dfrac{\tanh^{-1}(\frac{2(x-2)}{\sqrt{5}\sqrt{x^2-2x+5}})}{2\sqrt{5}} + C$

244) $-\dfrac{2\left(\sqrt{9\sqrt{5}-8}\,\tan^{-1}(\frac{\sqrt{6x+1}}{\sqrt{8+9\sqrt{5}}}) + \sqrt{8+9\sqrt{5}}\,\tanh^{-1}(\frac{\sqrt{6x+1}}{\sqrt{9\sqrt{5}-8}})\right)}{3\sqrt{1705}} + C$

245) $\dfrac{52}{3} \approx 17.333$

246) $\dfrac{1}{2}(27\sqrt[3]{3} - 1) \approx 18.970$

247) $\dfrac{1}{8}(8 + \pi^2) \approx 2.2337$

248) $1 - \dfrac{\pi^2}{8} \approx -0.2337$

249) $\dfrac{171\pi}{2} \approx 268.61$

250) 0

251) $\ln(49) \approx 3.8918$

252) $\dfrac{1106}{3} \approx 368.67$

253) $\dfrac{1}{3}(56 + 13\sqrt{13}) \approx 34.291$

254) $\dfrac{1}{e^2} + e^3 + 2\left(\ln(256) - 6\right) \approx 19.311$

255) $\dfrac{40}{3} \approx 13.333$

256) $\dfrac{26}{3} + 32\sqrt{3} \approx 64.0923$

257) 240

258. $\dfrac{\pi \ln(2)}{8}$

259. $\dfrac{\pi}{4}$

260. $\dfrac{1}{2}$

261. $\dfrac{\pi}{4}$

262. π

263. $\dfrac{\pi^2}{4}$

264. 0

265. $\dfrac{\pi^2}{6}$

266. $32/5$

267. π

268) $\dfrac{1}{2}\left(\dfrac{x\sqrt{x^2+16}}{16}+\dfrac{1}{2}\left(\ln\left(\dfrac{\pi}{4}+\sqrt{1+\dfrac{x^2}{16}}\right)-\ln\left(-\dfrac{\pi}{4}+\sqrt{1+\dfrac{x^2}{16}}\right)\right)\right)$

269. $\dfrac{61030\pi}{3}$

270) $\ln(x^2+2x+10)+\dfrac{1}{3}\tan^{-1}(\dfrac{x+1}{3})+C$

271) $\dfrac{1}{2}\left(25\sin^{-1}(\dfrac{x}{5})-x\sqrt{25-x^2}\right)+C$

272) $\dfrac{2}{231}\sin^{\frac{3}{2}}(x)\left(21\sin^4(x)-66\sin^2(x)+77\right)+C$

274) $\ln\left(\sin(x) + \cos(x)\right) + C$

275) ≈ 1.11072

276) $\frac{1}{6}\left(-e^{2x} + \ln(e^{2x} + 1) + 2e^{3x}\tan^{-1}(e^x)\right) + C$

277) $\frac{1}{2}\sqrt{(x+1)^2 + 5}(x+1) + \frac{5}{2}\sinh^{-1}(\frac{x+1}{\sqrt{5}}) + C$

278) $\frac{1}{3}\left(\ln(2-x) - \ln(x+1)\right) + C$

279) $\frac{1}{30}\left((15-4\sqrt{15})\ln(-x+\sqrt{15}-2) + (15+4\sqrt{15})\ln(x+\sqrt{15}+2)\right) + C$

280) $2\tan^{-1}(\sqrt{x}) + C$

281) $\sin^{-1}(\frac{\sin(x)}{2}) + C$

282) $(x+1)\ln(x+1) - x + C$

283) $2\tanh^{-1}\left(\sin(\sqrt{x})\right) + C$

284) $-10\sin^{-1}\left(\sqrt{1-\dfrac{x}{2}}\right)+C$

285) $\dfrac{1}{10}\left((5+\sqrt{5})\ln(-2x+\sqrt{5}-1)-(\sqrt{5}-5)\ln(2x+\sqrt{5}+1)\right)+C$

287) $\ln\left(\sin(x)+\cos(x)\right)-\ln\left(\sin(x)+2\cos(x)\right)+C$

288) $\dfrac{2\sqrt{x}}{\sqrt{3}}-\dfrac{2}{3}\ln(\sqrt{3}\sqrt{x}+1)+C$

289) $x(x^2+x+3)+\dfrac{3}{2}\ln(1-x)-\dfrac{15}{2}\ln(x+1)+C$

290) $-\sqrt{-(x-6)x}-16\sin^{-1}(\sqrt{1-\dfrac{x}{6}})+C$

291) $\dfrac{1}{4}\left(\sec^4(x)-4\sec^2(x)-4\ln(\cos[x])\right)+C$

292) $\frac{1}{2}(x+3)\sqrt{x^2+6x+1}-4\ln(\sqrt{x^2+6x+1}+x+3)+C$

293) $\frac{1}{2}\left(\sqrt{x^2+6x+18}(x+3)+9\sinh^{-1}(\frac{x+3}{3})\right)+C$

294) $\frac{1}{192}\left(12x+3\sin(2x)-3\sin(4x)-\sin(6x)\right)+C$

295) $\dfrac{8\pi}{15}\approx 1.6755$

297) $\dfrac{1}{x}\cos(x^3) - \dfrac{3x^2}{2}\sin(x^3) + \dfrac{1}{2}\sin(x^3) + \dfrac{4}{x} + C$

298) $\dfrac{2}{3}\Big(2\ln(2-x) + \ln(x+1)\Big) + C$

299) $\dfrac{4^x}{\ln(4)} - \dfrac{2^{x+1}5^x}{\ln(10)} + \dfrac{25^x}{\ln(25)} + C$

300) $\dfrac{1}{60}\Big(-3\ln(2-x) + 2\ln(3-x) + 3\ln(x+2) - 2\ln(x+3)\Big) + C$

301) $\dfrac{1}{8}\ln(1-x^8) - \ln(x) + C$

302) $\dfrac{1}{6}\sqrt{3x^2+4x+5}\,(3x+2) + \dfrac{11\sinh^{-1}\left(\frac{3x+2}{\sqrt{11}}\right)}{6\sqrt{3}} + C$

303) $\dfrac{1}{4}\Big(2\sqrt{2x^2+4x+5}\,(x+1) + 3\sqrt{2}\sinh^{-1}(\sqrt{\tfrac{2}{3}}[x+1])\Big) + C$

304) $\dfrac{\tan^{-1}\left(\frac{3\tan(x)+1}{2\sqrt{x}}\right)}{2\sqrt{2}} + C$

305) $\dfrac{3-3x}{x^2}\ln(x) + C$

306) $x + e^x + \dfrac{e^{2x}}{2} + C$

307) $\dfrac{x}{2} + \sin(\dfrac{x}{2}) + C$

308) $\dfrac{1}{2}\ln(2x+3) + C$

309.

$$-\frac{\ln(x)+1}{x} + c$$

310.

$$\frac{1}{729}\left(-\frac{9}{x-6} - \frac{9}{x+3} - 2\ln|x-6| + 2\ln|x+3|\right) + c$$

311.

$$\frac{1}{33750}\left(\frac{15(2x^2-17x-67)}{(x-8)^2(x+7)} + 2\ln|x-8| - 2\ln|x+7|\right) + c$$

312.

$$-5\left(\sqrt{4x-x^2} + 4\arcsin\left(\sqrt{1-\frac{x}{4}}\right)\right) + c$$

313.

$$\frac{1}{2}x\left(x\sec^{-1}(x) - 1\sqrt{1-\frac{1}{x^2}}\right) + c$$

314.

$$\frac{1}{5544}\sin^7(x)(364\cos(2x) + 63\cos(4x) + 365) + c$$

315.

$$\frac{1}{6}\left(-\ln|x^2+x+1| + 2\ln|1-x| - 2\sqrt{3}\arctan\left(\frac{2x-1}{\sqrt{3}}\right)\right) + c$$

316.

$$-\frac{1}{2}\sqrt{-x^2+2x+1}(x+1) - \frac{\sqrt{2}}{2}\arctan\left(\frac{x}{\sqrt{-\frac{x^2}{2}+x+\frac{1}{2}}}\right) - 2\arcsin\left(\frac{1-x}{\sqrt{2}}\right) + c$$

317.

$$\ln|1-x| + 2\ln|x| - 3\ln|x+1| + c$$

318.

$$x\sec(x) + \ln\left|\cos\left(\frac{x}{2}\right) - \sin\left(\frac{x}{2}\right)\right| - \ln\left|\sin\left(\frac{x}{2}\right) + \cos\left(\frac{x}{2}\right)\right| + c$$

319.

$$\frac{1}{4}(2\sin x + \sin(2\sin x)) + c$$

320.

$$\frac{240}{\ln(3)}$$

321.

$$\frac{1}{2}(e^x \sqrt{e^{2x}+1} + \sinh^{-1}(e^x)) + c$$

322.

$$\frac{1}{1296}(432x^3 \arctan(6x) - 36x^2 + \ln|36x^2+1|) + c$$

323.

$$4\sqrt{x^2+6x-11} - 13\ln|\sqrt{x^2+6x-11} + x + 3| + c$$

324.

$$\frac{3}{2}\ln(x^2+1) + c$$

325.

$$\frac{\ln 2}{4} + \tanh^{-1}(\tan(\frac{\pi}{8}))$$

326) $\dfrac{7}{2}\tan^{-1}(e^{x^2}) + C$

327) $e^x(4x + 3) + C$

328) $\dfrac{3}{12}\left(2\sqrt{\cos(2x)}(\cos[x]-6\cos[x])+7\sqrt{2}\ln(\sqrt{2}\cos[x]+\sqrt{\cos[2x]})\right)+C$

329) 0

330) $\dfrac{1}{2}\left(1 + e\sin(1) - e\cos(1)\right) \approx 0.909033$

331) $-\dfrac{2}{45}(2 - x^3)^{\frac{3}{2}}(3x^3 + 4) + C$

332) $\dfrac{1}{2}\sqrt{x^2 + 2x + 5}(x + 1) + 2\sinh^{-1}(\dfrac{x+1}{2}) + C$

333) $\dfrac{1}{2}\tan^{-1}(\dfrac{x+1}{2}) + C$

334) $\dfrac{1}{2}\tan^{-1}(\dfrac{x+1}{2}) + C$

335) $(x - 1)x + \dfrac{5}{2}\ln(1 - x) + \dfrac{3}{2}\ln(x + 1) + C$

336) $\dfrac{765}{2} = 382.5$

337) $-\dfrac{7\sqrt{-(x-6)x}}{3x} + C$

338) $-\dfrac{7}{3}\ln(x^3 - 3) + C$

339) $\dfrac{3}{2}x(x+1) - \ln(x) + C$

340) $10\sqrt{x^2 + 1} + C$

341) $\dfrac{2}{\sqrt{2}}\tan^{-1}\left(\dfrac{\sqrt{3}\tan(\frac{x}{2}) + 1}{\sqrt{2}}\right) + C$

342) $\dfrac{28}{3} \approx 9.333$

344) $\dfrac{1 - 5x}{20(x-1)^5} + C$

345) $\dfrac{2^{3x^2+6x+1}}{\ln(8)} + C$

346) $\dfrac{1}{8}\left(2x^2 + \sin(2[x^2 + 1]) + 2\right) + C$

347) $\dfrac{1}{192}\left(\dfrac{-30x^4 - 90x^3 - 70x^2 + 30x + 64}{(x-1)(x+1)^4} - 15\ln(x-1) + 15\ln(x+1)\right) + C$

348) $4\left(\sqrt{1 - e^{2x}} + e^x \sin^{-1}(e^x)\right) + C$

349) $\dfrac{1}{4}\tan^{-1}\left(\dfrac{1}{2}(x^2 + 2)\right) + C$

350) $12\tan^{-1}(\sqrt{x}) + C$

351) $8\sqrt{2}\ln(\sqrt{2}\sqrt{x} + 1) + C$

352) $5\ln\left(\sin(x) + 4\cos(x)\right) - 5\ln\left(\cos(x)\right) + C$

353) $\dfrac{1}{2}\left(\sqrt{x^2 + 91}x + 91\sinh^{-1}(\dfrac{x}{\sqrt{91}})\right) + C$

354) $\dfrac{1}{4}\tan^{-1}(e^{4x}) + C$

355) $-4\sqrt{-x^2 - 15} + C$

356) $\frac{1}{128}\Big(3(8x^2-1)\sin(4x)-4x(8x^2-3)\cos(4x)\Big)+C$

357) $\frac{1}{3}\sqrt{2x+1}(2x+7)+C$

358) $\frac{2}{3}\Big(2\ln(4-\sin[x])+\ln(\sin[x]+2)\Big)+C$

359) $-\frac{x^2}{8}+\frac{1}{128}\ln(16x^2+1)+C$

360) $\frac{1}{3}\Big(x^3\ln(\frac{1-x}{x+1})-x^2-\ln(1-x^2)\Big)+C$

361) $\dfrac{2(2x+1)}{3\sqrt{x^2+x+1}}+C$

362) $\frac{1}{6}\Big(\sqrt{x^2-2x+2}(2x^2-x+1)-3\sinh^{-1}(1-x)\Big)+C$

363) $4\sqrt{x^2+6x-11}-13\ln(\sqrt{x^2+6x-11}+x+3)+C$

365) $\dfrac{x^2}{2}+\dfrac{10\tan^{-1}(\frac{4x+3}{\sqrt{23}})}{\sqrt{23}}+C$

366) $\dfrac{x-6}{2\sqrt{x^2-4}}+C$

367) $2\Big(\ln(x-2)-\dfrac{2}{x-2}\Big)+C$

368) $-4x^2 + \frac{4\ln(\sqrt[3]{3}x^2+3^{\frac{2}{3}}x+3)}{\sqrt[3]{3}} - \frac{8\ln(3-3^{\frac{2}{3}}x)}{\sqrt[3]{3}} - 8\sqrt[6]{3}\tan^{-1}\left(\frac{2x}{3^{\frac{5}{6}}} + \frac{1}{\sqrt{3}}\right) + C$

369) 69

370) $\frac{3}{8}\ln\left|4x^2 - 4x + 17\right| + \frac{1}{16}\arctan\left(\frac{1}{4}(2x-1)\right) + C$

371)
$$\int \frac{\sin(x) - 2\cos(x)}{1 + 4\sin(x)\cos(x)}\,dx = \frac{1}{12\sqrt{2}}$$
$$\left(18\tanh^{-1}\left((\sqrt{2}-\sqrt{3})\tan\left(\frac{x}{2}\right) + \sqrt{2}\right) + 18\tanh^{-1}\left(\frac{(2+\sqrt{6})\tan(\frac{x}{2}) + 2}{\sqrt{2}}\right) + \right.$$
$$\sqrt{3}\left(\log\left(-\sec^2\left(\frac{x}{2}\right)(-\sqrt{2}\sin(x) + \sqrt{2}\cos(x) + \sqrt{3})\right) - \right.$$
$$\left.\left.\log\left(-\sec^2\left(\frac{x}{2}\right)(2\sin(x) - 2\cos(x) + \sqrt{6})\right)\right)\right) + \text{constant}$$

372) $\frac{e^{7x}\sin(x)}{50} + \frac{7e^{7x}\cos(x)}{50} + C$

373) $\frac{7e^x\sin(7x)}{50} + \frac{e^x\cos(7x)}{50} + C$

374) $\frac{1}{3}\left(\tan^{-1}\left(\frac{x}{1-x^2}\right) + 2\tan^{-1}(x)\right) + \text{constant}$

375) $-\frac{9}{385}\ln|3x - 1| - \frac{1}{126}\ln|x + 2| + \frac{1}{594}\ln|x - 4| + \frac{4}{135}\ln|2x + 1| + C$

376) $-\frac{\arctan(x)}{2(x-1)^2} + \frac{1}{2}\left(\frac{1}{4}\ln\left|x^2 + 1\right| - \frac{1}{2}\ln|x - 1| - \frac{1}{2(x-1)}\right) + C$

377) $\frac{25^x}{2\ln(5)} - \frac{5^x \cdot 2^{x+1}}{\ln(10)} + \frac{2^{2x-1}}{\ln(2)} + C$

378) $-\frac{1}{6}\cot(6x) + C$

379) $-x^2\ln(1+x) - x\ln(1+x) + \frac{1}{3}\ln(1+x)(x+1)^3 - \frac{1}{9}(1+x)^3 + \frac{1}{2}(1+x)^2 - 1 - x + C$

218

381) $2\left(-\sqrt{2x+1}+\sqrt{3x+4}+\sqrt{5}\,\tan^{-1}\!\left(\frac{\sqrt{2x+1}}{\sqrt{5}}\right)-\sqrt{5}\,\tan^{-1}\!\left(\frac{\sqrt{3x+4}}{\sqrt{5}}\right)\right)+$
constant

382) $-\dfrac{4}{\sqrt{55}}\arctan\!\left(\sqrt{\dfrac{5}{11}}\,\cot(x)\right)+C$

383) $\dfrac{1}{\sqrt{10}}\arctan\!\left(\dfrac{x+3}{\sqrt{10}}\right)+C$

384) $\ln\left|\dfrac{1}{\sqrt{10}}(x+3)+\sqrt{\dfrac{1}{10}\left(x^2+6x+19\right)}\right|+C$

385) $\frac{1}{8}\left(-13\left(\ln\left|2x+\sqrt{13}\,\sqrt{\frac{4x^2+12x-4}{13}}+3\right|-\frac{\ln(13)}{2}\right)+2(2x+3)\sqrt{x^2+3x-1}\right)+C$

386) $2\sin^4(x)+C$

387) $-6e^{6+\cos(x)}+C$

388) $\dfrac{1}{2}\ln|6x+7|+C$

389) $\dfrac{1}{24}\left(6x^2+1-\ln\left|6x^2+1\right|\right)+C$

390) $\dfrac{1}{3}\ln|x-1|+\dfrac{2}{3}\ln|x+2|+C$

391) $\arcsin\!\left(\dfrac{1}{9}x\right)-\sqrt{81-x^2}+C$

392) $2\left(-\cos(3x)+\dfrac{1}{3}\cos^3(3x)\right)+C$

393) $4.81047\ldots$

395) $\dfrac{4\sqrt{2}-2}{3}+\dfrac{2}{3}$

396) $\frac{1}{4}\left(-2x^4\cos(2x)+4x^3\sin(2x)-3\left(-2x^2\cos(2x)+2x\sin(2x)+\cos(2x)\right)\right)+C$

397) $\dfrac{x^2}{2}+C$

398) $\displaystyle\int_0^{\frac{\pi}{4}}\dfrac{\cos(x)+\sin(x)}{9+16\sin(2x)}\,dx=0.0549306$

399) $\quad -\dfrac{\arctan(3x)}{x} + 3\ln(3|x|) - \dfrac{3}{2}\ln\left|9x^2 + 1\right| + C$

400) $\quad x - 3x^2 + \dfrac{11x^3}{3} - \dfrac{3x^4}{2} + C$

401) $\quad \dfrac{1}{97(1-x)^{97}} - \dfrac{1}{49(1-x)^{98}} + \dfrac{1}{99(1-x)^{99}} + C$

402) $\quad \dfrac{1}{2}\arcsin\left(\dfrac{1}{\sqrt{3}}\sin^2(x)\right) + C$

403) $\quad \dfrac{x^2}{2} - 2x - 18\left(-\dfrac{1}{9}\ln|x| + \dfrac{1}{3x} + \dfrac{1}{9}\ln|x-3|\right) + C$

404) $\quad \displaystyle\int_0^1 \sin^2(\log(x))\,dx = \dfrac{2}{5} = 0.4$

406) $\quad -\ln|x| + \dfrac{5}{6}\ln|3x-2| + \dfrac{1}{2}\ln|x+2| + C$

407) $\quad \dfrac{3}{2}\left(\dfrac{1}{2}\ln\left|36x^2 - 72x + 9\right| - \dfrac{5}{6\sqrt{3}}\left(\ln\left|\dfrac{2(x-1)}{\sqrt{3}} + 1\right| - \ln\left|\dfrac{2(x-1)}{\sqrt{3}} - 1\right|\right)\right) + C$

408) $\quad \dfrac{1}{2}\arctan\left(\dfrac{x+1}{2}\right) + C$

409) $\quad 0$

410) $\quad -\dfrac{1}{2}\cos(2x) + C$

411) $\quad \dfrac{10x^2 + 4x - x^4}{4e^{2x}} + \dfrac{1}{4}e^{-2x}x^4 + \dfrac{1}{2}e^{-2x}x^3 - \dfrac{7}{4}e^{-2x}x^2 - \dfrac{11}{4}e^{-2x}x - \dfrac{11}{8}e^{-2x} + C$

412) $\quad 4\left(x\left(-\dfrac{1}{2}e^x\cos(x) + \dfrac{1}{2}e^x\sin(x)\right) + \dfrac{1}{2}e^x\cos(x)\right) + C$

413) $\quad \dfrac{9}{16}\left(2\arcsin\left(\dfrac{2}{3}x - 1\right) + \sin\left(2\arcsin\left(\dfrac{2}{3}x - 1\right)\right)\right) + C$

414) $\int_0^{\frac{\pi}{2}} \dfrac{\cos^2(x)}{\cos^2(x) + 4\sin^2(x)}\, dx = \dfrac{\pi}{6} \approx 0.52360$

415) $-\dfrac{3}{x-2} + \ln\left|x^2 + 4x + 6\right| + C$

416) $\log(x^2 + 4x + 6) - \dfrac{3}{x-2} - \dfrac{\tan^{-1}\left(\frac{x+2}{\sqrt{2}}\right)}{\sqrt{2}} + \text{constant}$

418) $\frac{1}{4}\left(x\sqrt{1+x^2} + \ln\left|x + \sqrt{1+x^2}\right| + x\sqrt{x^2-1} - \ln\left|x + \sqrt{x^2-1}\right|\right) + C$

419) $-\frac{1}{2}\arctan(x) - \frac{1}{4}\ln|x+1| + \frac{1}{4}\ln|x-1| + C$

420) $\dfrac{(x^n + 1)^2}{2n} - x^n - 1 + C$

421) $\frac{8}{25}\left(\frac{1}{3}(5x+1)^{\frac{3}{2}} - \sqrt{5x+1}\right) + C$

422) $-8\cos(x) + 30\sin^2(x) - 50\cos^3(x) - \dfrac{125}{4}\cos^4(x) + C$

423) $\dfrac{\pi}{2}\ln\left(\dfrac{\pi}{2}\right)$

424) $\int_0^\infty \dfrac{\log(x)}{1+x^2}\, dx = 0$

425) $\dfrac{\pi}{2}$

426) $2\pi \ln(2)$

427. 0.94608

428. 0.44841

429. 0.29709

430. 0.16889

431. 0.79660

432. − 0.16189

433) $\quad -\dfrac{\pi^2}{4}$

434) $\quad \displaystyle\int_0^\infty \log(1 + e^{-x})\, dx = \dfrac{\pi^2}{12} \approx 0.822467$

435) $\quad \displaystyle\int_0^\infty \log\!\left(\dfrac{2^x - 1}{2^x}\right) dx = -\dfrac{\pi^2}{\log(64)}$;

436) $\quad \operatorname{Li}_2(i) = -\dfrac{\pi^2}{48} + i \displaystyle\sum_{k=0}^{\infty} \dfrac{(-1)^k}{(1 + 2k)^2}$

438) $\quad \dfrac{1}{2}i\Big(\operatorname{Li}_2(-i3x) - \operatorname{Li}_2(i3x)\Big) + C$

Made in the USA
Las Vegas, NV
27 November 2024

12833836R00125